Reviews of Environmental Contamination and Toxicology

VOLUME 164

Springer
*New York
Berlin
Heidelberg
Barcelona
Hong Kong
London
Milan
Paris
Singapore
Tokyo*

Reviews of Environmental Contamination and Toxicology

Continuation of Residue Reviews

Editor
George W. Ware

Editorial Board
Lilia A. Albert, Xalapa, Veracruz, Mexico
F. Bro-Rasmussen, Lyngby, Denmark · D.G. Crosby, Davis, California, USA
Pim de Voogt, Amsterdam, The Netherlands · H. Frehse, Leverkusen-Bayerwerk, Germany
O. Hutzinger, Bayreuth, Germany · Foster L. Mayer, Gulf Breeze, Florida, USA
N.N. Melnikov, Moscow, Russia · D.P. Morgan, Cedar Rapids, Iowa, USA
Douglas L. Park, Baton Rouge, Louisiana, USA
Annette E. Pipe, Burnaby, British Columbia, Canada
Raymond S.H. Yang, Fort Collins, Colorado, USA

Founding Editor
Francis A. Gunther

VOLUME 164

Springer

Coordinating Board of Editors

DR. GEORGE W. WARE, *Editor*
Reviews of Environmental Contamination and Toxicology

5794 E. Camino del Celador
Tucson, Arizona 85750, USA
(520) 299-3735 (phone and FAX)

DR. HERBERT N. NIGG, *Editor*
Bulletin of Environmental Contamination and Toxicology

University of Florida
700 Experimental Station Road
Lake Alfred, Florida 33850, USA
(941) 956-1151; FAX (941) 956-4631

DR. DANIEL R. DOERGE, *Editor*
Archives of Environmental Contamination and Toxicology

6022 Southwind Drive
N. Little Rock, Arkansas, 72118, USA
(501) 791-3555; FAX (501) 791-2499

Springer-Verlag
New York: 175 Fifth Avenue, New York, NY 10010, USA
Heidelberg: Postfach 10 52 80, 69042 Heidelberg, Germany

Library of Congress Catalog Card Number 62-18595.
Printed in the United States of America.

ISSN 0179-5953

Printed on acid-free paper.

© 2000 by Springer-Verlag New York, Inc.
All rights reserved. This work may not be translated or copied in whole or in part without the written permission of the publisher (Springer-Verlag New York, Inc., 175 Fifth Avenue, New York, NY 10010, USA), except for brief excerpts in connection with reviews or scholarly analysis. Use in connection with any form of information storage and retrieval, electronic adaptation, computer software, or by similar or dissimilar methodology now known or hereafter developed is forbidden.
The use of general descriptive names, trade names, trademarks, etc., in this publication, even if the former are not especially identified, is not to be taken as a sign that such names, as understood by the Trade Marks and Merchandise Marks Act, may accordingly be used freely by anyone.

ISBN 0-387-98927-7 Springer-Verlag New York Berlin Heidelberg   SPIN 10743147

Foreword

International concern in scientific, industrial, and governmental communities over traces of xenobiotics in foods and in both abiotic and biotic environments has justified the present triumvirate of specialized publications in this field: comprehensive reviews, rapidly published research papers and progress reports, and archival documentations. These three international publications are integrated and scheduled to provide the coherency essential for nonduplicative and current progress in a field as dynamic and complex as environmental contamination and toxicology. This series is reserved exclusively for the diversified literature on "toxic" chemicals in our food, our feeds, our homes, recreational and working surroundings, our domestic animals, our wildlife and ourselves. Tremendous efforts worldwide have been mobilized to evaluate the nature, presence, magnitude, fate, and toxicology of the chemicals loosed upon the earth. Among the sequelae of this broad new emphasis is an undeniable need for an articulated set of authoritative publications, where one can find the latest important world literature produced by these emerging areas of science together with documentation of pertinent ancillary legislation.

Research directors and legislative or administrative advisers do not have the time to scan the escalating number of technical publications that may contain articles important to current responsibility. Rather, these individuals need the background provided by detailed reviews and the assurance that the latest information is made available to them, all with minimal literature searching. Similarly, the scientist assigned or attracted to a new problem is required to glean all literature pertinent to the task, to publish new developments or important new experimental details quickly, to inform others of findings that might alter their own efforts, and eventually to publish all his/her supporting data and conclusions for archival purposes.

In the fields of environmental contamination and toxicology, the sum of these concerns and responsibilities is decisively addressed by the uniform, encompassing, and timely publication format of the Springer-Verlag (Heidelberg and New York) triumvirate:

Reviews of Environmental Contamination and Toxicology [Vol. 1 through 97 (1962–1986) as Residue Reviews] for detailed review articles concerned with any aspects of chemical contaminants, including pesticides, in the total environment with toxicological considerations and consequences.

Bulletin of Environmental Contamination and Toxicology (Vol. 1 in 1966) for rapid publication of short reports of significant advances and discoveries in the fields of air, soil, water, and food contamination and pollution as well as methodology and other disciplines concerned with the introduction, presence, and effects of toxicants in the total environment.

Archives of Environmental Contamination and Toxicology (Vol.1 in 1973) for important complete articles emphasizing and describing original experimental or theoretical research work pertaining to the scientific aspects of chemical contaminants in the environment.

Manuscripts for *Reviews* and the *Archives* are in identical formats and are peer reviewed by scientists in the field for adequacy and value; manuscripts for the *Bulletin* are also reviewed, but are published by photo-offset from camera-ready copy to provide the latest results with minimum delay. The individual editors of these three publications comprise the joint Coordinating Board of Editors with referral within the Board of manuscripts submitted to one publication but deemed by major emphasis or length more suitable for one of the others.

Coordinating Board of Editors

Preface

Thanks to our news media, today's lay person may be familiar with such environmental topics as ozone depletion, global warming, greenhouse effect, nuclear and toxic waste disposal, massive marine oil spills, acid rain resulting from atmospheric SO_2 and NO_x, contamination of the marine commons, deforestation, radioactive leaks from nuclear power generators, free chlorine and CFC (chlorofluorocarbon) effects on the ozone layer, mad cow disease, pesticide residues in foods, green chemistry or green technology, volatile organic compounds (VOCs), hormone- or endocrine-disrupting chemicals, declining sperm counts, and immune system suppression by pesticides, just to cite a few. Some of the more current, and perhaps less familiar, additions include *xenobiotic transport, solute transport, Tiers 1 and 2, USEPA to cabinet status, and zero-discharge*. These are only the most prevalent topics of national interest. In more localized settings, residents are faced with leaking underground fuel tanks, movement of nitrates and industrial solvents into groundwater, air pollution and "stay-indoors" alerts in our major cities, radon seepage into homes, poor indoor air quality, chemical spills from overturned railroad tank cars, suspected health effects from living near high-voltage transmission lines, and food contamination by "flesh-eating" bacteria and other fungal or bacterial toxins.

It should then come as no surprise that the '90s generation is the first of mankind to have become afflicted with *chemophobia*, the pervasive and acute fear of chemicals.

There is abundant evidence, however, that virtually all organic chemicals are degraded or dissipated in our not-so-fragile environment, despite efforts by environmental ethicists and the media to persuade us otherwise. However, for most scientists involved in environmental contaminant reduction, there is indeed room for improvement in all spheres.

Environmentalism is the newest global political force, resulting in the emergence of multi-national consortia to control pollution and the evolution of the environmental ethic. Will the new politics of the 21st century be a consortium of technologists and environmentalists or a progressive confrontation? These matters are of genuine concern to governmental agencies and legislative bodies around the world, for many serious chemical incidents have resulted from accidents and improper use.

For those who make the decisions about how our planet is managed, there is an ongoing need for continual surveillance and intelligent controls to avoid endangering the environment, the public health, and wildlife. Ensuring safety-in-use of the many chemicals involved in our highly industrialized culture is a dynamic challenge, for the old, established materials are continually being displaced by newly developed molecules more acceptable to federal and state regulatory agencies, public health officials, and environmentalists.

Adequate safety-in-use evaluations of all chemicals persistent in our air, foodstuffs, and drinking water are not simple matters, and they incorporate the judgments of many individuals highly trained in a variety of complex biological, chemical, food technological, medical, pharmacological, and toxicological disciplines.

Reviews of Environmental Contamination and Toxicology continues to serve as an integrating factor both in focusing attention on those matters requiring further study and in collating for variously trained readers current knowledge in specific important areas involved with chemical contaminants in the total environment. Previous volumes of *Reviews* illustrate these objectives.

Because manuscripts are published in the order in which they are received in final form, it may seem that some important aspects of analytical chemistry, bioaccumulation, biochemistry, human and animal medicine, legislation, pharmacology, physiology, regulation, and toxicology have been neglected at times. However, these apparent omissions are recognized, and pertinent manuscripts are in preparation. The field is so very large and the interests in it are so varied that the Editor and the Editorial Board earnestly solicit authors and suggestions of underrepresented topics to make this international book series yet more useful and worthwhile.

Reviews of Environmental Contamination and Toxicology attempts to provide concise, critical reviews of timely advances, philosophy, and significant areas of accomplished or needed endeavor in the total field of xenobiotics in any segment of the environment, as well as toxicological implications. These reviews can be either general or specific, but properly they may lie in the domains of analytical chemistry and its methodology, biochemistry, human and animal medicine, legislation, pharmacology, physiology, regulation, and toxicology. Certain affairs in food technology concerned specifically with pesticide and other food-additive problems are also appropriate subjects.

Justification for the preparation of any review for this book series is that it deals with some aspect of the many real problems arising from the presence of any foreign chemical in our surroundings. Thus, manuscripts may encompass case studies from any country. Added plant or animal pest-control chemicals or their metabolites that may persist into food and animal feeds are within this scope. Food additives (substances deliberately added to foods for flavor, odor, appearance, and preservation, as well as those inadvertently added during manufacture, packing, distribution, and storage) are also considered suitable review material. Additionally, chemical contamination in any manner of air, water, soil, or plant or animal life is within these objectives and their purview.

Normally, manuscripts are contributed by invitation, but suggested topics are welcome. Preliminary communication with the Editor is recommended before volunteered review manuscripts are submitted.

Tucson, Arizona G.W.W.

Table of Contents

Foreword v
Preface vii

Metalaxyl: Persistence, Degradation, Metabolism, and
Analytical Methods 1
    Premasis Sukul, and Michael Spiteller

Indoor Household Pesticides: Hazardous Waste Concern or Not? 27
    John M. Owens, Patrick D. Guiney, Philip H. Howard,
    Dallas B. Aronson, and D. Anthony Gray

Mercury Modeling to Predict Contamination and Bioaccumulation in
Aquatic Ecosystems 69
    M.C.B. Braga, G. Shaw, and J.N. Lester

Biomarkers in Terrestrial Invertebrates For Ecotoxicological Soil
Risk Assessment 93
    Jan E. Kammenga, Reinhard Dallinger, Marianne H.
    Donker, Heinz R. Köhler, Vibeke Simonsen, Rita Triebskorn,
    and Jason M. Weeks

Index 149

Metalaxyl: Persistence, Degradation, Metabolism, and Analytical Methods

Premasis Sukul · Michael Spiteller

Contents

I. Introduction	1
II. Abiotic Degradation	4
A. Effect of Heat	4
B. Hydrolysis	5
C. Photodegradation	5
III. Biotic Degradation and Metabolism	6
A. Plant	6
B. Soil	6
C. Rat	9
IV. Analytical Methods	9
A. Thin Layer Chromatography (TLC)	11
B. Gas Liquid Chromatography (GLC)	11
C. High Performance Liquid Chromatography (HPLC)	14
D. Micellar Electrokinetic Capillary Chromatography (MEKC)	14
E. Mass Spectrometry (MS)	15
F. Bioassay	15
G. Enzyme-Linked Immunosorbent Assay (ELISA)	15
V. Uptake, Translocation, Persistence, and Degradation	16
VI. Adsorption and Mobility	18
Summary	19
Acknowledgment	20
References	20

I. Introduction

Metalaxyl [*N*-(2,6-dimethylphenyl)-*N*-(methoxyacetyl)alanine methyl ester] is effective against Oomycetes, especially Peronosporales such as *Phytophthora* spp., *Pseudoperonospora* spp., *Peronospora* spp., *Plasmopara* spp., *Sclerospora* spp., *Bremia* spp., *Pythium* spp., and other species causing downy mildews, late

---

Communicated by George W. Ware

P. Sukul
Department of Agricultural Chemistry and Soil Science, Faculty of Agriculture, Bidhan Chandra Krishi Viswavidyalaya, Mohanpur 741252, India

M. Spiteller (✉)
University of Dortmund, Institute of Environmental Research, Otto-Hahn-Str. 6, 44221 Dortmund, Germany.

blight, damping off, and root, stem, and fruit rots (Schwinn et al. 1977; Urech et al. 1977; Kerkenaar and Sijpestejin 1981; Houseworth 1987) in crops such as pearl millet (Venugopal and Safeeulla 1978; Williams and Singh 1981; Dang et al. 1983; Siddiqui et al. 1987; Pandya et al. 1994), sugarcane (Malein 1993), capsicum (Matheron and Matejka 1995), mustard and rapeseed (Dueck and Stone 1979; Sharma and Kolte 1985), sunflower (Sackston 1979), maize, bazra, and sorghum (Urech et al. 1978), tobacco (Bruin et al. 1981), and tomato (Cohen et al. 1979). Several reports are available regarding its systemic properties in different crops and showing the relation between fungicide accumulation in plants during growth and disease control (Kotwal et al. 1981; Tonini and Avigliano 1981). It is a systemic fungicide and is mainly translocated upward in plants, although some amount of downward translocation has also been reported (Tripathi and Singh 1983). Because of its broad-spectrum activity, metalaxyl is registered for use in many countries worldwide including the United States, the European nations, Australia, and India on a variety of fruit and vegetable crops. Information about its general characteristics is given in Table 1, and maximum residue limits (MRL) on different commodities, as fixed by WHO (1991), are listed in Table 2.

Metalaxyl is a stable compound, resistant to a wide range of pH, temperature, and light (Singh and Tripathi 1982). These properties lead to its abundant use in agriculture. Its spectrum of activity is broadened to control more diseases when it is used with other fungicides such as dithiocarbamates, phthalimides, or copper fungicides and at the same time the buildup of resistant fungal strains may be delayed or even prevented (Businelli et al. 1984; Fontem and Aighewi 1993). It is used as seed treatment and as a root and foliar fungicide (Schwinn 1983). Its *R*-enantiomeric form is more biologically active than the *S*-form. The *R*-form provides the same level of efficacy as the mixture but at half the application rate. Therefore, the introduction of the *R*-form (CGA 329351, mefenoxam) may contribute toward risk reduction for a compound with an excellent safety profile (Nuninger et al. 1996).

Studies on the mode of action of metalaxyl revealed specific interference in certain steps of the infection process and in the metabolic pathway of pathogens. Release, mobility, encystment, and germination of zoospores, e.g., in *Phytophthora infestans* and *Plasmopara viticola*, as well as initial penetration and primary haustorium development, are relatively insensitive to metalaxyl. On the other hand, the compound effectively inhibited further development of the pathogens beyond primary haustorium formation. Incorporation studies revealed that synthesis of lipids, proteins, and DNA was either barely or to a lesser degree affected by metalaxyl than was incorporation of uridine into RNA of *Pythium* and *Phytophthora* species. Further investigations with *Phytophthora megasperma* f.sp. *medicaginis* indicated that metalaxyl did not inhibit labeled uridine uptake and conversion of uridine into uridine triphosphate (UTP), suggesting that metalaxyl inhibited uridine incorporation into RNA by interfering at the transcriptional level. Metalaxyl specifically inhibits RNA polymerase-1 of the fungi belonging to the Peronosporales, and blocking of rRNA synthesis may be

Table 1. General characteristics of metalaxyl.

1. Chemical class	Acylalanine
2. Chemical name	Methyl N-(2-methoxyacetyl)-N-(2,6-xylyl)-DL-alaninate
3. Molecular formula	$C_{15}H_{21}NO_4$
4. Molecular weight	279.34
5. Enentiomer mixture [R:S, %]	50:50
6. Chemical structure	$H_3CO-\overset{O}{\overset{\|}{C}}-\overset{CH_3}{\underset{H}{\overset{\|}{C}}}-N-\overset{O}{\overset{\|}{C}}-CH_2OCH_3$ with 2,6-dimethylphenyl group on N
7. Trade name	Ridomil, Ridomil Plus, Apron 35 WS, Apron 70 SD, Fubol, Acylon Super F, CGA 48988
8. Manufacturer	Novartis Crop Protection AG
9. Crystal density	1.21 g/ml at 20 °C
10. Melting point	71–72 °C
11. Vapor pressure	0.75 mPa (25 °C)
12. Octanol–water partition coefficient (log P_{ow})	1.75
13. λ_{max}	205.3 nm in methanol, 196 nm in aqueous solution
14. Solubility (mg/L)	In water: 8.4 g/L at pH 5.2 and 22 °C (pure water) In organic solvents at 25 °C: n-hexane 11 g/L; toluene 340 g/L; dichloromethane 770/g/L; ethanol 400/g/L; n-octanol 68 g/L; acetone 450 g/L
15. Stability	Stable to 300 °C; stable in neutral and acidic media at room temperature; on hydrolysis, DT_{50} (calculated) (20 °C) >200 d at pH 1, 115 d at pH 9, 12 d at pH 10
16. Toxicology	Acute oral LD_{50} for rats: 669 mg/kg body weight Acute dermal LD_{50} for rats: >3100 mg/kg body weight Inhalation LC_{50}/4 hr for rats: >3600 mg/m^3 Eye irritation rabbits: slight Skin irritation rabbits: slight Sensitization in guineapigs: not sensitizing Wildlife organisms: practically nontoxic to fish, birds, bees

Table 1. (Continued).

	NOEL for rats: 2.5 mg/kg body weight daily
	NOEL for mice: 31.7 mg/kg body weight daily
	NOEL for dogs: 8.0 mg/kg body weight daily
	ADI (JMPR): 0.03 mg/kg body weight
	Toxicity class: WHO III
17. Antidote	No specific antidote is known
	Symptomatic therapy is recommended

Source: Singh and Tripathi 1982; Banerjee 1995; Tomlin 1995.

regarded as the primary mode of action of metalaxyl (Buchenauer 1990). Metalaxyl is reported to influence the nutrient status and amount of biomolecules present in crops (Chakravarti et al. 1997), possibly because of its ability to influence soil microorganisms. It is evident that the stimulatory response of VA mycorrhizal fungi to low concentrations of metalaxyl resulted in increased plant biomass production, nutrient uptake, and grain yield of wheat (Shetty and Magu 1997).

II. Abiotic Degradation
A. Effect of Heat

Metalaxyl is stable to heat. Autoclaving had no effect on the stability of metalaxyl (Singh and Tripathi 1982), which was unaltered both qualitatively and quantitatively. Toxicity was also unchanged as LD_{50} values calculated for both treated and untreated metalaxyl remained the same. This property would make in vitro studies easier, particularly in bioassay work, because it can be added into the medium before sterilization, minimizing chances of contamination.

Table 2. Maximum residue limits (MRL) in foods or commodities.

Food or commodity	MRL (mg/kg)
Apple, asparagus, cereal grains, cotton seed, pea (shelled), potato, soybean (dry), sugarbeet, sunflower seed	0.05
Peanut	0.10
Avocado, melon, summer and winter squash	0.20
Tomato	0.50
Grapes, peppers	1.0
Citrus fruits	5.0
Hops, dry	10.0

Source: WHO (1991).

B. Hydrolysis

Metalaxyl is very stable in water within a pH range of 1.0 to 8.5 (Singh and Tripathi 1982; Sharom and Edgington 1982). The fungicide degrades rapidly in water at pH 10.0, with less than 5% of the initial amount remaining after 12 wk (Sharom and Edgington 1982). At 20 °C, the calculated half-life was 200 d at pH 5 and 7 and 115 d at pH 9 (EPA 1994).

C. Photodegradation

One of the most important pathways in abiotic degradation of pesticides is photodegradation. For its photoreaction, a compound must absorb light energy directly or indirectly. Compounds with no ultraviolet (UV) absorption above 290 nm appear to be safe from photochemical breakdown because the ozone layer in the upper atmosphere absorbs all radiation emitted by the sun below 290 nm (Watkins 1979). A distinction must be made between direct photolysis, initiated by direct absorption of light by the chemical, and indirect or sensitized photolysis, involving light absorption by natural photosensitizers. Only light with a wavelength greater than 290 nm reaches the surface of the earth. The UV absorption spectrum of a compound can thus give basic information about the possibility of direct photodegradation, because only substances with an UV absorption in the range of sunlight can undergo direct interaction. In fact, most pesticides have UV absorption bands that fade out in the border area of 280–290 nm or below, and so practically no direct interaction with sunlight is possible. In contrast, humic substances can strongly absorb in the UV region of sunlight and can therefore act as sensitizers and initiators of degradation processes in nonabsorbing chemicals. Metalaxyl is resistant to sunlight because its λ_{max} is 196 nm in aqueous solution and no absorption occurs above 290 nm.

Photolysis of metalaxyl in UV light leads to rearrangement of the *N*-acyl group to the aromatic ring (Yao et al. 1989), demethoxylation, *N*-deacylation, and elimination of the methoxycarbonyl group from the molecule (Sukul et al. 1992). An acid metabolite with 5% yield is also reported after irradiation of metalaxyl aqueous solution for 7 d using artificial sunlight (Burkhard 1979).

Sukul et al. (1992) reported irradiation of the fungicide at 254 nm, which resulted in 53% attenuation in 3 hr. Long-time (65-hr) irradiation under simulated light (Suntest apparatus, Heraeus, Germany) in the presence of commercially available humic acid resulted in 65% degradation of metalaxyl, while no degradation was observed without humic acid. Humic acid is an important ingredient of soil, and soil has been reported to produce active oxygen species (Gohre and Miller 1983; Smith et al. 1978; Slawinski et al. 1978). Humic acid produces active oxygen species (OH radical), which degrade metalaxyl.

Usually pH does not affect the stability of metalaxyl (Singh and Tripathi 1982; Sharom and Edgington 1982), but in a separate experiment Yao et al. (1989) found that pH affects its rate of transformation when it is irradiated with UV light in aqueous solution. Exposure to UV light in aqueous solutions of pH 2.8–8.8 resulted in more efficient photodegradation of metalaxyl at lower pH values.

Photosensitizers such as (acetone, riboflavin, rhodamime-B, humic acid, TiO_2, and H_2O_2) accelerated photodecomposition of metalaxyl (Singh and Tripathi 1982; Yao et al. 1989; Sukul et al. 1992; Moza et al. 1994). Irradiation in the presence of H_2O_2 results in oxidation of the molecule, and in the presence of TiO_2 deacylation takes place (Moza et al. 1994). H_2O_2 and TiO_2 under UV light and water produce OH radicals (Dorfman and Adams 1973; Oliver and Corey 1986) that enhance the degradation of agrochemicals. Several compounds with photosensitizing properties are naturally present inside the plant cells or are released outside in leaf exudates (Towers 1980). Therefore, metalaxyl is photometabolized on the plant surface.

Metalaxyl on oven-dried soil photodecomposes slowly (Murthy et al. 1998) as compared to UV-induced transformation in water (Yao et al. 1989; Sukul et al. 1992) and on moist soil (Saha and Sukul 1997). Less than 2% conversion of metalaxyl on oven-dried soil was found when it was subjected to artificial sunlight for 72 hr (Murthy et al. 1998). 2,6-Dimethyl aniline (in waterlogged soil), 2,6-dimethyl-*N*-ethylacetanilide (air-dried soil), and *N*-(2,6-dimethylphenyl)-alanine methyl ester (in both air-dried and waterlogged soil) were found when metalaxyl was irradiated on soil in natural sunlight under different moisture status (Saha and Sukul 1997). These three metabolites were also isolated and identified when the fungicide was subjected to UV irradiation in aqueous medium (Sukul et al. 1992). The photoproducts obtained on irradiation of metalaxyl in aqueous solution are shown in Fig. 1.

III. Biotic Degradation and Metabolism
A. Plant

Studies with potatoes, lettuce, grapes, and tobacco indicate that metalaxyl is taken up, translocated, and extensively metabolized by plants (Businelli et al. 1984; Gross 1986; Owen and Donzel 1986; Cole and Owen 1987; EPA 1994). Metabolism involves oxidation of a ring-methyl group and hydrolysis of the methyl ester and methyl ether bonds; metabolites can be conjugated to glucose. Ring-methyl hydroxylation was found predominantly in cell suspension culture of lettuce and grapevine (Cole and Owen 1987). Other prominent products arose from *O*-dealkylation, further ester hydrolysis of *O*-dealkylated product, or ester hydrolysis only. Minor residues resulted from combined *O*- and *N*-dealkylation and aryl hydroxylation (Fig. 2). Species differences are also evident in metalaxyl metabolism. It has been observed that ester hydrolysis is less important in grapevine whereas *N*-dealkylation and aryl hydroxylation are less important in lettuce (Owen and Donzel 1986; Cole and Owen 1987). In whole lettuce plants, the same metabolites were identified (Gross 1986) as in cell suspension culture, although in another study only the ester hydrolysis product (acid metabolite) was identified in lettuce and sunflower (Businelli et al. 1984).

B. Soil

The main breakdown product of metalaxyl in soil is the acid metabolite (Droby and Coffey 1991) (Fig. 3). However, in another study the degradation of metal-

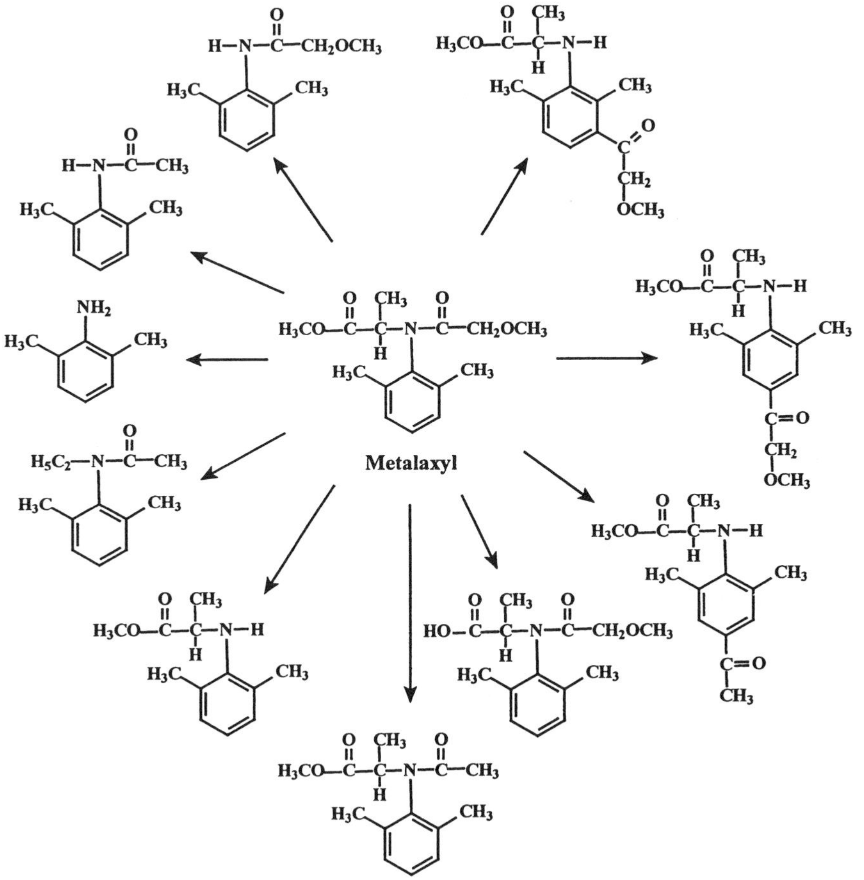

Fig. 1. Photoproducts following irradiation of metalaxyl in aqueous solution (from Burkhard 1979; Yao et al. 1989; Sukul et al. 1992).

axyl in soil was mainly found by benzylic hydroxylation of the methyl side chain and biodegradation (Wang et al. 1995). Metalaxyl is relatively stable in sterilized loam and muck soils but is degraded by native microflora with a half-life of 3 and 8 wk, respectively, in unsterilized soil; 90% of the applied metalaxyl remained in sterile soil even after 12 wk (Sharom and Edgington 1982). Soils with previous exposure to metalaxyl were found to degrade the fungicide faster (enhanced biodegradation) than similar soils with no history of metalaxyl exposure (Bailey and Coffey 1985; Droby and Coffey 1991). Investigation of the microbial transformation of metalaxyl by the fungus *Syncephalastrum racemosum* has revealed that the major transformation mechanism involves benzylic or aromatic ring hydroxylation (Zheng et al. 1989). Therefore, these hydroxylated compounds may also be expected in soils. However, Zheng et al. (1989) were unable to detect metabolites similar to those observed by Bailey

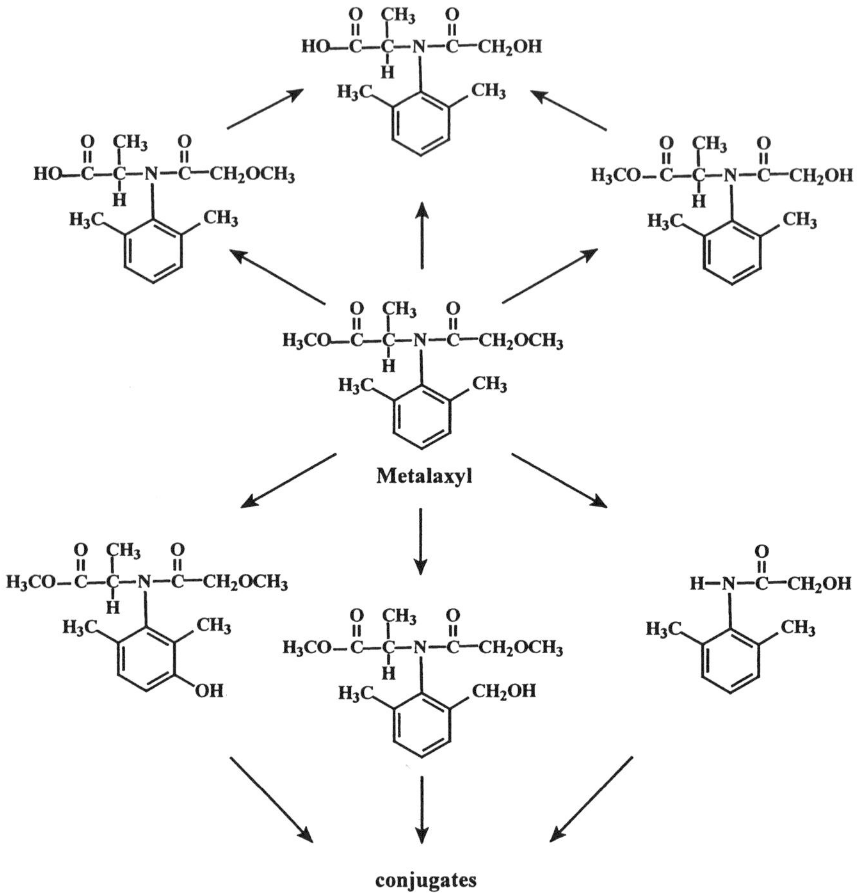

Fig. 2. Metabolic breakdown of metalaxyl in whole plant and tissue culture (from Gross 1986; Cole and Owen 1987).

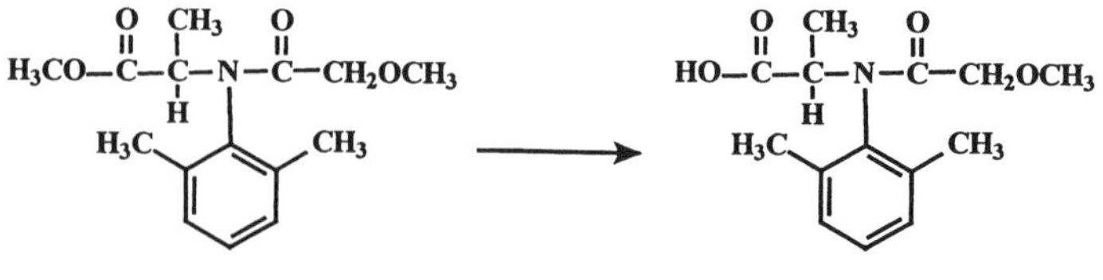

Fig. 3. Breakdown of metalaxyl to its acid metabolite in soil.

and Coffey (1986). This result indicates that the metabolism of metalaxyl in synthetic media by certain soil microorganisms may not accurately reflect the process of degradation taking place under natural soil conditions wherein a broad spectrum of organisms is probably involved.

In general, the aerobic soil metabolism half-life of metalaxyl was determined to be about 40 d. The major degradate, CGA-62826 (*N*-2,6-dimethylphenyl-*N*-methoxyacetyl-alanine) was further broken down to nonextractable material and CO_2. This acid metabolite accounted for as much as 53.6% of the applied material at 66 d and thereafter degraded to 23% at 360 d. At 12 mon, metalaxyl accounted for less than 2% of the applied material and nonextractable residues accounted for 38.3% (EPA 1994).

C. Rat

Oral doses of metalaxyl (27.8 mg/kg) to female rats resulted in 58% of the radioactivity being excreted in the urine and 32% in the feces within 48 hr. Many of the metabolites excreted were in both their free forms and their glucuronide conjugates. It has been suggested that the degradation of metalaxyl in the rats proceeds primarily via (1) hydrolysis of methyl ester and methyl ether groups, (2) oxidation of the 2(6)-methyl group, (3) oxidation of the phenyl ring, and (4) *N*-dealkylation and subsequent formation of glucuronic acid conjugates (Hambock 1978, 1981). A metabolic pathway for metalaxyl degradation in the rat is proposed in Fig. 4.

In a comprehensive study (EPA 1994), metalaxyl pharmacokinetics with male and female Sprague-Dawley rats following a single intravenous dose (1 mg/kg), single oral low dose (1 mg/kg), single oral high dose (200 mg/kg), or repeated oral doses (1 mg/kg/d for 14 d) have been evaluated. For female rats, urine is the major elimination route; in the males, it is the feces. Three major and one minor metabolic pathways were proposed. One pathway involved hydrolysis of the ether, followed by oxidation of the resulting alcohol, ester hydrolysis, or *N*-dealkylation of ester chain. A second pathway involved oxidation of an aromatic methyl to the benzylic acid or ester hydrolysis. The third major pathway was ester hydrolysis, with sometimes hydroxylation at the meta-position of the phenyl ring.

IV. Analytical Methods

Analytical methods are required for routine monitoring of pesticides in crop matrices as well as in different environmental compartments. A number of TLC, GLC, HPLC, and other methods are reported for metalaxyl quantification, which are discussed separately here.

In different foods, soils, or other commodities, metalaxyl and its metabolites have been determined both quantitatively and qualitatively after extraction with acetone (Singh and Tripathi 1980; Caverly and Unwin 1981; Stone at al. 1987; Getzin et al. 1989; Mehta et al. 1997), acetonitrile (Tafuri et al. 1981; Murthy

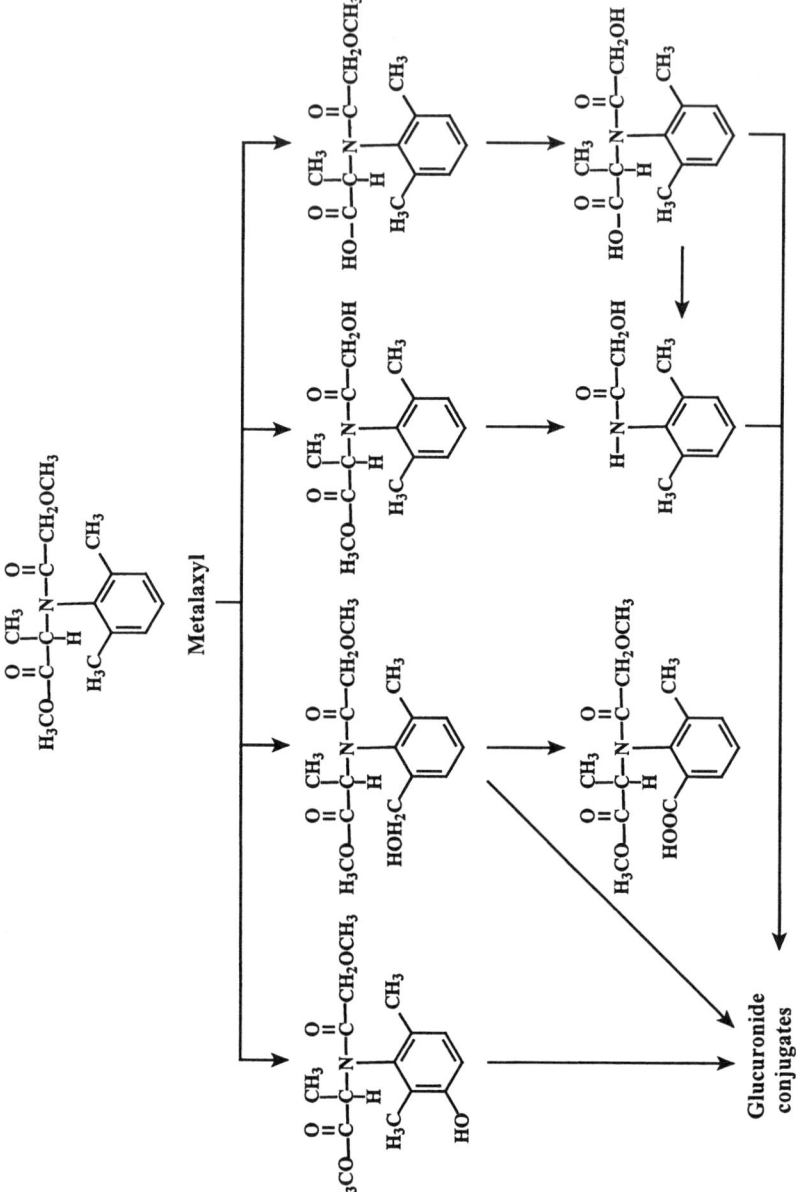

Fig. 4. Metabolic pathways proposed for the degradation of metalaxyl in the rat (from Hambock 1978, 1981).

et al. 1998), methanol (Wynn and Crute 1983; Newsome 1985; Khazanchi et al. 1989; Saha and Sukul 1997), or dichloromethane (Singh and Tripathi 1980; FAO/WHO 1996). For cleanup, in addition to different conventional liquid partition and adsorption chromatography, sweep-codistillation to separate metalaxyl from nonvolatile compounds followed by silica gel column chromatography is also reported (Tafuri et al. 1981). Solid-phase extraction (SPE) cartridges, consisting of a solid-phase silicon bed with attached 18-carbon chain (C_{18}) functional groups, may also be used for isolation of metalaxyl from a water matrix (Novak and Watts 1997). C_{18}-SPE cartridges require less solvent and time and are capable of replacing many of the tedious liquid-liquid extraction procedures for various chemicals in water samples.

A. Thin Layer Chromatography (TLC)

Benzene methanol (150/5, v/v), benzene ethylacetate (9/8, v/v), or carbon tetrachloride ethylacetate (10/9, v/v) as the developing solvent mixture and iodine azide, iodine potassium iodide, or iodine azide followed by iodine potassium iodide as chromogenic reagents may be successfully used to detect very low levels (2.5 µg) of metalaxyl (Singh and Tripathi 1980). Iodine azide produces brown spots on a light yellow background while iodine potassium iodide generates dark brown spots on a yellowish-brown background, but the quick disappearance of the color of the spots is a disadvantage in the use of these two chromogenic reagents individually. Use of iodine azide followed by iodine potassium iodide is much preferred as it gives a stable spot on a light brown background. This method has been used successfully to analyze metalaxyl from treated maize and pearl millet plants, giving a recovery of about 95% (Singh and Tripathi 1980), in residue analysis in tobacco (Bhatt et al. 1987), and in photodecomposition studies (Singh and Tripathi 1982). Yao et al. (1989) used TLC more conveniently to isolate photodecomposition products on glass plates associated with 0.25-mm silica gel with a solvent system benzene acetone (9/1, v/v). Spots were visualized under UV light.

B. Gas Liquid Chromatography (GLC)

With the help of GLC, metalaxyl has been quantitatively estimated by two methods: after extraction and cleanup steps, the fungicide can be measured directly (operating parameters are listed in Table 3); and indirectly (Balasubramanian and Perez 1982; Reddy et al. 1994). An improved derivatization method of metalaxyl determination has been developed (Balasubramanian and Perez 1982) in which metalaxyl is converted to 2,6-dimethylaniline by refluxing the crop extractant with methanesulfonic acid, estimated using a NP detector operating in nitrogen-specific mode at 300 °C. A capillary column SE-54 (0.2 mm × 25 m) or DX-4 (0.32 mm × 30 m) was used. The limit of detection was 0.05 ppm expressed as metalaxyl equivalents. Metalaxyl most often has been measured by gas chromatography using a nitrogen-selective detector (Caverly and Unwin 1981; Businelli et al. 1984; Vuik et al. 1989; Seefeld and Dunsing 1990; Tafuri

Table 3. Gas chromatography (GC) parameters for metalaxyl determination.

Detector	Column	Operating parameters	Retention time (min)	Sensitivity	Reference
Alkali flame ionisation	5% OV-3 on chromosorb WHP (80–100 mesh)	Column temp. 240 °C Injection temp. 250 °C Detector temp. 300 °C Flow rate 30 mL He/min	2.5	0.1 mg/kg	Speck and Dirr 1980
Rubidium chloride thermoionic	5% high vacuum silicone grease on 80–100 mesh Gaschrom-Q	Column temp. 190 °C Detector temp. 250 °C Flow rate 40 mL N_2/min	—	0.02–0.5 mg/kg	Caverly and Unwin 1981
Nitrogen selective detector	1.5% cyclohexanedimethanol succinate on 80–100 mesh Gaschrom-Q	Column temp. 190 °C Injector temp. 240 °C Detector temp. 240 °C Flow rate 25 mL He/min	7.5	0.05 µg/ml	Tafuri et al. 1981
Thermoionic specific detector	Suppelcowax 10 megabore capillary column	Column temp. programmed for 220 °C to 250 °C with 10 °C/min increase Injector temp. 250 °C Detector temp. 275 °C Flow rate 30 mL He/min	5	0.02 µg/g	Businelli et al. 1984; Getzin et al. 1989
Flame ionization detector	Capillary column RT_x-5, fused silica tubing cross-bonded with 95% di-methyl polysiloxane-5% diphenyl polysiloxane	Column temp. 220 °C Programmed from 100 °C to 220 °C at a rate of 8 °C/min Injector temp. 250 °C Detector temp. 275 °C	—	—	Yao et al. 1989
NP detector	Capillary column of fused silica, CP-SIL5, and BP-10	Column temp. programmed from 100 °C to 180 °C at 20 °C/min and then 180 °C to 260 °C at 10 °C/min with 2-min hold at 260 °C carrier gas flow rate 3 mL He/min	—	0.04 mg/kg	Vuik et al. 1989

Table 3. (Continued).

Detector	Column	Operating parameters	Retention time (min)	Sensitivity	Reference
Alkali flame ionization detector with rubidium sulfate	5% carbowax on 60–80 mesh chromosorb P.	Column, Injector and detector temp. 170 °C, 210 °C, and 200 °C, respectively Flow rate 50 mL N_2/min	—	—	Reddy et al. 1990
NP detector	3% OV-275 on varaport-30 (80–100 mesh) or 1.95% OV-210 + 1.5% OV-17 on chromosorb W (HP) 80–100 mesh or a fused silica column coated with 50% phenyl methyl silicone	Column temp. 210 °C Nitrogen as carrier gas	—	—	Seefeld and Dusing 1990
Electron capture detector	SE-30 + OV-210	Injector, column, and detector temp. 230 °C, 280 °C, and 245 °C, respectively Nitrogen as carrier gas, 60 mL/min	—	—	Rao et al. 1993
NP detector	Capillary column RT_x-50, cross-bonded 50% phenyl-50% methyl polysiloxane	Injector and detector temp. 225 °C and 300 °C, respectively Column temp. programmed 175 °C–225 °C at 5 °C/min	7.5	4 ppb	Petrovic et al. 1996
Thermoionic specific detector	10% OV-101	Column, injector and detector temp. 230 °C, 240 °C, and 280 °C, respectively Nitrogen as carrier gas, 30 mL/min	2.8	—	Mohapatra et al. 1997

et al. 1981), alkali flame ionization detector (Speck and Dirr 1980; Reddy et al. 1990), thermoionic specific detector (Getzin et al,. 1989; Mohapatra et al. 1997); and flame ionization detector (Yao et al. 1989). Reports are also available on using a gas chromatograph equipped with an electron-capture detector (Rao et al. 1993; Reddy et al. 1994) to estimate metalaxyl residues in different crops.

C. High Performance Liquid Chromatography (HPLC)

HPLC is one of the most preferred analytical techniques used in pesticide analysis. The analysis of metalaxyl formulation and technical product has been carried out on HPLC using benzophenone as the internal standard (Khazanchi and Roy 1987). A Lichrosorb RP-8 column with a flow rate of 2 mL/min for a mobile-phase methanol–water (65:35) and an UV detector at 275 nm was used to detect metalaxyl. Later, Khazanchi et al. (1989) used the same method with a little variation in the mobile phase (methanol–water, 4:1), using a flow rate of 1 mL/min instead of 2 mL/min, for determination of metalaxyl in treated pearl millet seeds. A programmed flow of mobile phase to measure the fungicide and its biodegradation products using C-18 (Vydac 201 HS) column and an UV detector operating at 274 nm is reported (Droby and Coffey 1991). For the first 5 min the mobile phase (1.5 mL/min) was composed of 85% acetonitrile and 15% water; for the next 20 min the acetonitrile content was decreased to 70%. To overcome the problem of separation of different metabolites of metalaxyl, a buffered solvent system may also be used successfully (Sukul et al. 1992). For that purpose, methanol–water (60:40), adjusted to pH 4.0 with 0.15 M H_3PO_4 and flushed with helium, was used as the mobile phase (0.5 mL/min) in a RP-18 column.

D. Micellar Electrokinetic Capillary Chromatography (MEKC)

Capillary electrophoresis (CE) provides a powerful new analytical tool for rapid and simultaneous determination of fungicide residues in surface water and groundwater samples and is complementary to GC and HPLC, offering many advantages over the conventional techniques. Application of CE for analysis of agrochemicals is gaining in popularity. Capillary electrophoresis with ultraviolet detection (CE/UV) of metalaxyl in combination with three other fungicides—carbendazim, propiconazol, and vinclozolin—using different buffer systems was investigated (Penmetsa et al. 1997). The four fungicides were well resolved by employing MEKC. Among the two surfactants tested in MEKC, bile salts provided better separation than SDS. A buffer consisting of 10 mM sodium phosphate with 100 mM sodium cholate and 10% methanol (pH 7.0) gave the best results. Separation of the target compound was carried out in less than 15 min. Average recovery of metalaxyl in lake water at 4 µg/L level was 87%.

E. Mass Spectrometry (MS)

Mass spectra of metalaxyl and its metabolites may be obtained on a GC-MS at an ionization potential of 70 eV (Yao et al. 1989; Sukul et al. 1992; Moza et al. 1994; Saha and Sukul 1997; Murthy et al. 1998). The GC conditions are as follows: a Machery-Nagel capillary column (Permabond SE-52; 25 m × 0.35 mm i.d.) coated with a 0.5-µm film of phenyl ethyl silicone: injection temperature, 230 °C; carrier gas, helium; temperature program, 70 °–250 °C, 10 °C/min.

Headley et al. (1996) reported a GC-MS method for the determination of combined residues of metalaxyl and its metabolites in urine containing the 2,6-dimethylaniline moiety. The method is a modification of a method of Balasubramanian and Perez (1982) for the analysis of metalaxyl in vegetables. Noted modifications include replacement of steam extraction with extraction by methylene chloride and use of electron impact ionization GC/MS in the selected ion mode. The method is linear over the range of 0.1 to 5 µg 2,6-dimethylaniline/g urine and has a detection limit of 0.025 mg/g.

Confirmation of the identity of metalaxyl was also carried out by GC-MS in the electron impact mode and LC thermospray MS-MS using both positive and negative ion modes and filament on (Salau et al. 1994).

F. Bioassay

Reports on a sensitive bioassay method for the quantification of metalaxyl in soil (Bailey and Coffey 1984) and in plant tissues (Lazorovits et al. 1982; Wynn and Crute 1983; Milgroom and Fry 1986; Van Bruggen et al. 1987) are available. A limitation of the bioassay method is its inability to differentiate between metalaxyl and any of its metabolites, or other toxicants with fungistatic activity against test organisms. These methods are suitable only where facilities for physicochemical methods are unavailable.

G. Enzyme-Linked Immunosorbent Assay (ELISA)

ELISA is a modern technique having immense prospects in pesticide analysis. An alternative approach to residue analysis with the potential of more efficient processing of samples is immunochemical determination based on competitive binding to an antibody (Hammock and Mumma 1980). ELISA has the advantage of not requiring a radioligand and associated counting equipment. Estimation of metalaxyl residues in various food commodities by ELISA technique is also possible (Newsome 1985). Methanol extracts without prior cleanup were analyzed and quantitated from 0.1 to 2.0 ppm in various commodities such as cucumber, squash, avocado, tomato, and potato. The method is quite comparable with the GLC method, which normally involves solvent partitioning and adsorption column cleanup. The coefficient of variation of six samples of tomato at 5 ppm level was 4.2% by ELISA and 5.9% by GLC, indicating a similar degree of within-run reproducibility by the two methods. The simplicity of the method permits 4.5 times more samples to be analyzed per day than with a conventional GC method. Good reproducibility and overall good results were also obtained during determination of metalaxyl residues on tobacco by the ELISA technique (Coussirat and Jacob 1992) However, because of its lack of specificity, this method is suitable only for screening and confirmation.

V. Uptake, Translocation, Persistence, and Degradation

Lipophilicity, measured as octanol–water coefficient, log P_{ow} (Edgington 1981) is one of the important parameters that control the adsorption, mobility, and translocation of any compound into the plant system. The octanol–water coefficient of metalaxyl (log P_{ow} 1.75 at 25 °C) suggests that the fungicide should be adsorbed only to a lower degree and translocated readily. Metalaxyl is readily taken up by the roots of tomato (Cohen et al. 1979; Zaki et al. 1981; Mohapatra et al. 1997), cabbage, red raspberry, strawberry (Carris and Bristow 1987), potato (Rowe 1982), and avocado (Zaki et al. 1981) and translocated to other plant parts. The concentration remaining in the roots at any given time may be useful to control root diseases (Mohapatra et al. 1997). Metalaxyl has been found to accumulate in the leaf margin as well as in the leaf lamina (Wynn and Crute 1981; Zaki et al. 1981; Singh et al. 1985). Basipetal translocation of metalaxyl into roots is only minimal (<1% after 2 wk) in tomato (Zaki et al. 1981) and somewhat higher (7% after 24 hr) in pea (Singh et al. 1985). In a field study, the greatest accumulation of metalaxyl was observed in lower leaves of rape seedlings grown from metalaxyl-treated seeds (Stone et al. 1987). The fungicide dissipated more or less completely in 14 d in cowpea, 30–45 days in pearl millet, and 45 d in maize (Tripathi and Singh 1983; Reddy et al. 1990), whereas fungicide residues persisted in tomato leaves and roots even after 30 d and residues in tomato-cropped soil reduced at a half-life of 18 d to persist more than 30 days (Mohapatra et al. 1997). In a greenhouse experiment no residue was found in harvest samples of tomato (Cabras et al. 1985), but a 7-d safe waiting period was recommended between the treatment and harvest of tomato in a hydroponic culture with recirculating nutrient solution (Dunsing et al. 1988).

In a field trial, no fungicide residues were detected in grape leaves after 1 month of its application at 0.25 and 2.5 kg a.i./ha. However, metalaxyl residues of 0.004–0.04 and 0.04–0.087 ppm were found in wine and 0.008 and 0.015 ppm of fungicide in soil at the end of the growing season, respectively (Vasilieva et al. 1991). In another experiment, metalaxyl residues in wine samples in Italy were reported below tolerance, indicating that it is safe for use in viticulture (Scarponi and Martinetti 1992).

On a coastal sandy soil of Western Australia, calculated half-lives showed 70 d for metalaxyl (Kookana et al. 1995); average time for its 50% dissipation in the turf soil profile was 16 d under golf course fairway conditions (Horst et al. 1996). The pesticide appeared to be dissipated more rapidly from turf grass than from other agronomic crops. Bailey and Coffey (1985) reported the half-life of metaxyl in soil that had received repeated application of the fungicide over 2–5 yr to range from 14 to 28 d. Soil previously treated with metalaxyl subsequently shows faster rates of breakdown (enhanced biodegradation), as evidenced in a tobacco soil giving a 6-d half-life for the fungicide (Droby and Coffey 1991). In the same experiment, addition of antibiotics to the soil greatly inhibited the rate of decomposition. The repeated addition of metalaxyl to soil induces the buildup of metalaxyl-degrading microorganisms that could degrade the fungicide at a faster rate.

Soil moisture influences the proliferation of microorganims and their activities, sorption–desorption of organic and inorganic compounds, and aerobic status in the soil (Sethunathan and Siddaramappa 1978); this is also applicable to metalaxyl. Mohapatra and Awasthi (1997a) found that the enrichment cultures from flooded soil degraded metalaxyl faster than cultures from dry soil, but not so fast as cultures from irrigated soil. Complete degradation of metalaxyl by enrichment cultures from irrigated soil could result from the involvement of both aerobic and anaerobic soil microorganisms. Enhancement of metalaxyl degradation by synergistic interaction among fungal cultures isolated from soils was found in a mineral salt medium (Mohapatra and Awasthi 1997b).

Seeds collected from metalaxyl-treated pearly millet and mustard plants at harvest were free from residues (Reddy et al. 1990; Mehta et al. 1997), possibly because of metabolic decomposition inside plant tissues (see Fig. 2) as well as dilution from plant growth. In mustard, when the fungicide was applied as a foliar spray at a recommended dose, the residue dissipated more or less completely by 15 d after the first and second spray; fungicidal absorption in mustard plants (using treated seeds) increased up to 30 d after seeding and thereafter declined to reach a nondetectable level within 60 d after seeding (Mehta et al. 1997). This result suggests its persistence in the plant parts for a longer period when applied as a seed dressing. When metalaxyl is used as a seed treatment, a major portion of the fungicide is lost through diffusion from seed into the soil during germination or remains in seed parts that are later shed to the ground (Singh et al. 1986). It would be interesting to know whether the lost fungicide in soil is partly absorbed by plant roots.

A dose-dependent increased uptake of metalaxyl was noticed in pearl millet during its use as soil treatment (Reddy et al. 1990). In general, leaves maintained high levels of residues irrespective of the type of crop and nature of treatment (Zaki et al. 1981; Reddy et al. 1990; Mohapatra et al. 1997); this could result from a direct relationship between fungicide accumulation and transpiration water loss, as demonstrated by Peterson and Edgington (1971). In separate experiments, metalaxyl persisted in maize for 29 d following seed treatment (Cho 1981), for 4 wk in lettuce (Curte 1980), and for the entire crop season in pea and sunflower (Brokenshire 1980; Niklov 1981).

Van Bruggen et al. (1987) studied the attenuation of metalaxyl on potato leaves by simulated acidic rain; as in industrialized areas, the pH of precipitation has dropped considerably during the last few years (Rothert and Dana 1984). Metalaxyl concentration on potato leaves declined rapidly over time and with simulated rain, regardless of its acidity (pH 2.8–4.6). Despite attenuation to 5%–10% of the original metalaxyl concentration, potato late blight was controlled by the remaining fungicide when applied at the recommended dose. The concentration remaining after 3 d or after 4.4 mm of simulated rain (6–18 µg/g of dry leaf tissue) was similar to the earlier findings in lettuce (Wynn and Crute 1981). Metalaxyl concentration declined rapidly even without rainfall; only 25% of the original concentration remained 30 hr after spraying. A possible explanation for this dissipation would be translocation, or metabolism in plant tissue and volatilization at the leaf surface (Wynn and Crute 1981; Van Bruggen et al. 1987).

VI. Adsorption and Mobility

In unaged leaching studies, parent ^{14}C-metalaxyl leached rapidly in sandy soil with up to 92% of the radioactivity recovered in the leachate. In sandy clay loam and silty loam soils, the majority of the radioactivity was recovered in the 6 to 24-cm soil layers with less than 0.6% in the leachates. In aged leaching studies, 79.2% of the applied radioactivity was found in the leachate and 16.1% remained in the soil of the sandy soil column; 56% of the activity found in the leachate was parent, 31% was acid metabolite, and 12% was unidentified. In the silty loam soil, 48.7% was in the leachate and 34.9% remained in the column. Using radiolabeled metalaxyl, partition coefficient (K_d) values of 0.43, 0.48, 0.87, and 1.40 for sandy, sandy loam, silty loam, and sandy clay loam soil were observed (EPA 1994).

Metalaxyl appeared to be preferentially adsorbed on soil mineral surfaces (Sukop and Cogger 1992). Although the metalaxyl molecule has no charged sites, it probably is adsorbed on soil by hydrogen bonding through the R_3N^+- and O-C-R groups or through van der Waals bonding at the hydrophobic sites (Sharma and Awasthi 1997). Significant spatial variations in soil properties such as organic matter content, bulk density, and moisture content affect the mobility, persistence, and fate of organic pesticides in the soil.

In a separate study conducted on a coastal sandy soil of Western Australia to measure the leaching and degradation rate of metalaxyl, the mean leaching depth (MLD) of metalaxyl for a 5-mon period was found to be 18 cm with a calculated half-life of 70 d (Kookana et al. 1995). The K-values obtained from Freundlich adsorption isotherms were 0.04 µg/g for a loamy sand and 10.15 and 13.32 µg/g for two clay soils. Metalaxyl mobility in these soils was inversely related to K-values ($r = -0.97$) (Sharma and Awasthi 1997). The mobility of metalaxyl in four Ontario soils, as influenced by various amounts of simulated rainfall, was found to be inversely related to the adsorption capacity ($r = -0.83$) and organic matter ($r = -0.85$) content of the soils (Sharom and Edgington 1982), while low adsorption of metalaxyl in soils with low organic matter and clay content was reported by Bailey and Coffey (1984).

The mobility of metalaxyl under unsaturated flow condition is greatly influenced by the adsorption capacity of the soil. Metalaxyl persisted for more than

60 d and did not move vertically to a lower soil layers (17.5–27.5 cm) in any of the experimental soils under unsaturated flow condition (Sharma and Awasthi 1997). However, this is also true for saturated soils having more clay content (Sharom and Edgington 1982). Most of the applied metalaxyl was also found to be retained in the top 5 cm of soil (Aylmore et al. 1995; Horst et al. 1996).

Groundwater and drinking water monitoring studies demonstrate that metalaxyl and its acid metabolite have the potential to reach groundwater, although in other studies indicated that metalaxyl occurred in groundwater only rarely and at levels well below 0.1 µg/L (Egli 1998). Detectable concentrations of pesticides have been reported in groundwater (EPA 1991); thus, attempts are currently underway to analytically survey water supplies, groundwater, and runoff for pesticides used and to investigate the downward migration characteristics of pesticides. Because metalaxyl possesses high water solubility and is used on golf course trees and greens that are constructed on sand, knowledge of its leaching behavior in sandy media is important. Few reports are available on the leaching behavior of metalaxyl on golf courses (Odanaka et al. 1994; Petrovic et al. 1996). A lysimetric study showed that the presence of peat in the sand delayed the downward migration of metalaxyl and that the percentage of total applied fungicide in the collected leachates decreased as surface organic matter increased (Petrovic et al. 1996). In another lysimetric study conducted in Japan on a creeping bentgrass turf soil (a 5-cm mixed upper layer of 50% sand, 20% peat moss, and 30% clay with a 35-cm layer of sand below), metalaxyl was found to readily leach downward after application of 10 cm water (Odanaka et al. 1994). The experiment was conducted, however, in winter when uptake of applied water by the grass would be minimal. Low adsorption of metalaxyl in soil with low organic matter and clay content was also reported by Bailey and Coffey (1984).

Greater microbial activity in dense root zones with concomitant formation of humus and active root uptake of water and dissolved solutes would be expected to contribute to a reduced downward movement of metalaxyl because of losses by its adsorption on organic matter, biodegradation, and root adsorption (Petrovic et al. 1996)

Summary

Metalaxyl is a systemic fungicide used to control plant diseases caused by Oomycete fungi. Its formulations include granules, wettable powders, dusts, and emulsifiable concentrates. Application may be by foliar or soil incorporation, surface spraying (broadcast or band), drenching, and seed treatment. Metalaxyl registered products either contain metalaxyl as the sole active ingredient or are combined with other active ingredients (e.g., captan, mancozeb, copper compounds, carboxin). Due to its broad-spectrum activity, metalaxyl is used worldwide on a variety of fruit and vegetable crops. Its effectiveness results from inhibition of uridine incorporation into RNA and specific inhibition of RNA polymerase-1. Metalaxyl has both curative and systemic properties. Its mammalian toxicity is classified as EPA toxicity class III and it is also relatively nontoxic to most nontarget arthropod and vertebrate species.

Adequate analytical methods of TLC, GLC, HPLC, MS, and other techniques are available for identification and determination of metalaxyl residues and its metabolites. Available laboratory and field studies indicate that metalaxyl is stable to hydrolysis under normal environmental pH values. It is also photolytically stable in water and soil when exposed to natural sunlight. Its tolerance to a wide range of pH, light, and temperature leads to its continued use in agriculture. Metalaxyl is photodecomposed in UV light, and photoproducts are formed by rearrangement of the N-acyl group to the aromatic ring, demethoxylation, N-deacylation, and elimination of the methoxycarbonyl group from the molecule. Photosensitizers such as humic acid, TiO_2, H_2O_2, acetone, and riboflavin accelerate its photodecomposition.

Information is provided on the fate of metalaxyl in plant, soil, water, and animals. Major metabolic routes include hydrolysis of the methyl ester and methyl ether oxidation of the ring-methyl groups. The latter are precursors of conjugates in plants and animals. In soils the most relevant metabolite is the metalaxyl acid, which is formed predominantly by soil microorganisms. Plant uptake, microbial degradation, photodecomposition, and leaching are the major route of metalaxyl dissipation. It has a tendency to migrate to deeper soil horizons with a potential to contaminate groundwater, particularly in soils with low organic matter and clay content. Therefore, precautions should be taken for the continuous application of metalaxyl to crops. If use of metalaxyl is greately increased, the risk of occurrence in groundwater must be reassessed, as by monitoring studies in the most vulnerable areas in main use regions.

The R-isomer of metalaxyl (mefenoxam) has recently been registered as the only active compound. Therefore, quantitative studies on the fate of this specific isomer are needed, including appropriate analytical methods. As the use rates of mefenoxam are approximately one-half those recommended for metalaxyl and mefenoxam dissipates more rapidly, concerns for mefenoxam reaching groundwater are even less justified.

Acknowledgment

This study was funded as part of a 1-yr research grant to the first author from the Alexander von Humboldt Foundation, Germany. The authors thank Mr. A. Achermann and Dr. J. Amrein, Novartis Crop Protection AG, for proofreading the manuscript. The authors deeply appreciate the help rendered by Dr. U. Klaus and Mrs. C. Köpk, University of Kassel, during the preparation of the manuscript.

References

Aylmore LAG, Kookana RS, Di HG, Heatwole C (1995) Pesticide leaching and model evaluation under field conditions. Proceedings of the International Symposium on Water Quality Modelling, Orlando, FL, April 2–5, 1995, pp 28–136.

Bailey AM, Coffey MD (1984) A sensitive bioassay for the quantification of metalaxyl in soil. Phytopathology 74:667–669.
Bailey AM, Coffey MD (1985) Biodegradation of metalaxyl in avocado soils. Phytopathology 75:135–137.
Bailey AM, Coffey MD (1986) Characterisation of microorganisms involved in accelerated biodegradation of metalaxyl and metolachlor in soils. Can J Microbiol 32:562–569.
Balasubramaniam K, Perez R (1982) Improved method for the determination of total residues of metalaxyl in crops as 2,6-dimethyl aniline. Biochemistry Department, Agricultural Division, Ciba-Geigy Greensboro, North Carolina.
Banerjee K (1995) Persistence and metabolism of metalaxyl and mancozeb on pearl millet. Ph.D. thesis, Division of Agricultural Chemicals, Indian Agricultural Research Institute, New Delhi, India.
Bhatt PA, Patel DJ, Patel BK (1987) Metalaxyl residue in tobacco. Indian J Plant Prot 15:166–167.
Brokenshire T (1980) Control of pea downy mildew with seed treatments and foliar sprays. Ann Appl Biol Suppl 94:34–35.
Bruin GCA, Ripley RD, Edgington LV (1981) Efficacy for blue mold control and persistence on tobacco. Tobacco Sci 25:128–130.
Buchenauer H (1990) In: Bowers WS, Ebing W, Martin D, Wegler R (eds) Chemistry of Plant Protection: Part 6: Controlled Release, Biochemical Effects of Pesticides, Inhibition of Plant Pathogenic Fungi. Springer-Verlag, Heidelberg, p 235.
Burkhard N (1979) Photolysis of GCA 48988 (Ridomil) in aqueous solution under artificial sunlight conditions. Project Report: Agrochemicals Division, Ciby-Geigy, Basel, Switzerland.
Businelli M, Patumi M, Marucchini C (1984) Identification and determination of some metalaxyl degradation products in lettuce and sunflower. J Agric Food Chem 32:644.
Cabras P, Meloni M, Pivisi MF, Cabitza F (1985) Behaviour of acylanilide and dicarboximidic fungicide residues on green house tomatoes. J Agric Food Chem 33:86–89.
Carris LM, Bristow PR (1987) Absorption and translocation of metalaxyl in cabbage, red raspberry and strawberry. J Agric Food Chem 35:851–855.
Caverly DJ, Unwin J (1981) Determination of residues of fluraxyl and metalaxyl in nutrient solution, peat compost and soil samples by gas chromatography. Analyst 106:389–393.
Chakravarti Dipankar, Thakore BBL, Chakravarti D (1997) Effect of metalaxyl spray on nitrogen, phosphorus, potassium and chlorophyll content of cucurbits. J Mycol Plant Pathol 27:72–74.
Cho JJ (1981) *Phytophthora* rot of *Banksia*, its host range and chemical control. Plant Dis 65:83–88.
Cohen M, Reuveni M, Eyal H (1979) The systemic antifungal activity of ridomil against *Phytophthora infestans* on tomato plants. Phytopathology 69:645–649.
Cole DJ, Owen WJ (1987) Metabolism of metalaxyl in cell suspension cultures of *Lactuca sativa* L. and *Vitis vinifera* L. Pestic Biochem Physiol 28:354–361.
Coussirat JC, Jacob A (1992) Determination of metalaxyl by ELISA. Ann Tab 24:35–39.
Curte IR (1980) Evaluation of lettuce downy mildew with metalaxyl seed treatment. Ann Appl Biol Suppl 94:50–53.
Dang JK, Thakur DP, Grover RK (1983) Control of pearl millet downy mildew caused by *Sclerospora graminicola* with systemic fungicides in an artificially contaminated plot. Ann Appl Biol 102:99–106.

Dorfman LM, Adams GE (1973) National Bureau of Standards Rep NSRDS-NBS-46. U.S. Government Printing Office, Washington, DC.

Droby S, Coffey MD (1991) Biodegradation process and the nature of metabolism of metalaxyl in soil. Ann Appl Biol 118:543–553.

Dueck J, Stone JR (1979) Evaluation of fungicides for control of *Albugo candida* in turnip rape. Can J Plant Sci 59: 423–427.

Dunsing M, Grote D, Buesi L (1988) Research into the persistence and residue dynamics of metalaxyl in tomatoes in hydroponic culture with recirculating nutrient solution (NFT). Nachrichtenbl Dtsch Pflanzenschutzdienst (Berl) 42:173–176.

Edgington LV (1981) Structural requirements of systemic fungicides. Annu Rev Phytopathol 19:107–124.

Egli H (1998) Metalaxyl in groundwater: situation in Europe and conclusions for Mefenoxam. Internal report. Novartis Crop Protection AG, Basel, Switzerland.

EPA (1991) National survey of pesticides in drinking water wells, Phase 1. Available from National Technical Information Service, Springfield, VA.

EPA (1994) Reregistration eligibility decision (RED): Metalaxyl. EPA 738-R-94-017. U.S. Environmental Protection Agency, Washington, DC.

FAO/WHO (1996) Pesticide residues in food—1995. Part I—Residues. FAO Plant Production and Protection Paper 137, p. 507. FAO/WHO, Rome, Italy.

Fontem DA, Aighewi B (1993) Effect of fungicides on late blight control and yield loss of potato in the Western highlands of Cameroon. Int J Pest Manage 32:152–155.

Getzin LW, Cogger CG, Bristow PR (1989) Simultaneous gas chromatographic determination of carbofuran, metalaxyl and simazine in soils. J Assoc Off Anal Chem 72: 361–364.

Gohre K, Miller GC (1983) Singlet oxygen generation on soil surfaces. J Agric Food Chem 31:1104–1108.

Gross D (1986) Uptake, translocation and metabolism of metalaxyl in higher plants. 6th International Congress on Pesticide Chemistry, Ottawa, August 1986.

Hambock H (1978) Metabolism of CGA 48988 in the rat. Project Rep 26/78. Ciba-Geigy Ltd., Basel, Switzerland.

Hambock H (1981) Metabolic pathways of CGA 48988 in the rat. Project rep 31/81. Ciba-Geigy Ltd., Basel, Switzerland.

Hammock BD, Mumma RO (1980) In: Harvey J Jr, Zweig G (eds) Recent Advances in Pesticide Analytical Methodology. ACS Symp. Ser. 136. American Chemical Society, Washington, DC, pp 321–352.

Headley JV, Maxwell DB, Swyngedouw C, Purdy JR (1996) Determination of combined residues of metalaxyl and 2,6-dimethylaniline metabolites in urine by gas chromatography/mass spectrometry. J Assoc Off Anal Chem Int 79:117–123.

Horst, GL, Shea PJ, Christians N, Miller DR, Stuefer-Powell C, Starrett SK (1996) Pesticide dissipation under golf course fairway conditions. Crop Sci 36:362–370.

Houseworth LD (1987) Excerpts from the new products and services from industry. Plant Dis 71:286–288.

Kerkenaar A, Sijpestijn AK (1981) Antifungal activity of metalaxyl and fluraxyl. Pestic Biochem Physiol 15:71–78.

Khazanchi R, Roy NK (1987) Reversed-phase high performance liquid chromatography of metalaxyl in technical formulations and from soil. Orient J Chem 3:3.

Khazanchi R, Dikshit AK, Roy NK, Kumar J (1989) Determination of metalaxyl in treated Bajra seeds by high performance liquid chromatography. Int J Trop Agric 7: 129–133.

Kookana RS, Di HJ, Aylmore LAG (1995) A field study of leaching and degradation of nine pesticides in a sandy soil. Austr J Soil Res 33:1019–1030.

Kotwal I, Arora A, Jain AC, Vyas SC (1981) Downward translocation of Ridomil in soybean seedlings. Pesticides 15:36.

Lazorovits G, Brammall RA, Ward EWB (1982) Bioassay of fungitoxic compounds on thin layer chromatograms with *Pythium* and *Phytophthora* species. Phytopathology 72:61–63.

Malein PJ (1993) Fungicidal control of *Peronosclerospora sacchari* (T. Miyake) Shirai and K. Hara in sugarcane in Papua New Guinea. Int J Pest Manage 39:325–327.

Matheron ME, Matejka JC (1995) Comparative activities of sodium tetrathiocarbonate and metalaxyl on *Phytophthora capsici* and root and crown rot on chile pepper. Plant Dis 79:56–59.

Mehta N, Saharan GS, Kathpal TS, Mehta N (1997) Absorption and degradation of metalaxyl in mustard plant (*Brassica juncea*). Ecotoxicol Environ Saf 37:119–124.

Milgroom MG, Fry WE (1986) Bioassay for metalaxyl in potato leaf tissue. Phytopathology 76:656.

Mohapatra S, Awasthi MD (1997a) Degradation of metalaxyl by enrichment cultures from sandy loam soil developed under different moisture regimes. Pestic Res J 9:36–40.

Mohapatra S, Awasthi MD (1997b) Enhancement of metalaxyl degradation by synergistic interaction among bacterial and fungal isolates. Pestic Res J 9:62–66.

Mohapatra S, Awasthi MD, Ahuja DK (1997) Uptake and translocation of metalaxyl in tomato plants. Pestic Res J 9:32–35.

Moza PN, Sukul P, Hustert K, Kettrup A (1994) Photooxidation of metalaxyl in aqueous solution in the presence of hydrogen peroxide and titanium dioxide. Chemosphere 28:341–347.

Murthy NBK, Hustert K, Moza PN, Kettrup A (1998) Photodegradation of selected fungicides on soil. Fresenius Environ Bull 7:112–117.

Newsome WH (1985) An enzyme linked immunosorbent assay for metalaxyl in foods. J Agric Food Chem 33:528–530.

Niklov B (1981) Apron 35 SD, an effective preparation for the control of downy mildew of sunflower. Rastit Zasht 29:40–41.

Novak JM, Watts DW (1997) Evaluation of C_{18} solid-phase extraction cartridges for the isolation of select pesticides and metabolites. J Environ Sci Health B 32:565–581.

Nuninger C, Watson G, Leadbitter N, Ellgehausen H (1996) CGA 329351: introduction of the enantiomeric form of the fungicide metalaxyl. In: Proceedings of the British Crop Protection Conference: Pests and Diseases—1996, vol 1, Brighton, Nov. 18–21, 1996, pp 41–46.

Odanaka Y, Taniguchi T, Shimamura Y, Iijima K, Koma Y, Takechi T, Matano O (1994) Runoff and leaching of pesticides in golf course. J Pestic Sci 19:1–10.

Oliver BG, Corey JH (1986) In: Pelizzetti E, Serpone N. (eds) Homogeneous and Heterogeneous Photocatalysis. Reidel, Dortrecht, p 629.

Owen WJ, Donzel B (1986) Oxidative degradation of chlortoluron, propiconazole and metalaxyl in suspension cultures of various crop plants. Pestic Biochem Physiol 26:75.

Pandya RK, Bartaria AM, Sharma BL (1994) Efficacy of ridomil against downy mildew of pearl millet. Indian Phytopathol 47:339.

Penmetsa VK, Leidy RB, Shea D (1997) Separation of fungicides by micellar electrokinetic capillary chromatography. Electrophoresis 18:235–240.

Peterson CA, Edgington LV (1971) Transport of benomyl into various plant organs. Phytopathology 61:91.
Petrovic AM, Barrett WC, Kovach IML, Reid CM, Lisk DJ (1996) The influence of a peat amendment and turf density on downward migration of metalaxyl fungicide in creeping bentgrass sand lysimeters. Chemosphere 33:2335–2340.
Rao BN, Reddy KN Sultan MA (1993) Residues of metalaxyl MZ and mancozeb in grape berries. Indian J Plant Prot 21:211–213.
Reddy KN, Rao BN, Sultan MA (1994) Residues of metalaxyl MZ and mancozeb in/on betelvine. Pestic Res J 6:84–86.
Reddy MVB, Shetty HS, Reddy MS (1990) Mobility, distribution and persistence of metalaxyl residues in pearl millet (*Pennisetum americanum* (L.) Leeke). Bull Environ Contam Toxicol 45:250–257.
Rothert JE, Dana MT (1984) The MAP3S. Precipitation Chemistry Network: 7th Periodic Summary Report (1983). Pacific Northwest Laboratory, Richland, WA.
Rowe RC (1982) Translocation of metalaxyl and RE 26745 in potato and comparison of foliar and soil application for control of *Phytophthora infestans*. Plant Dis 66: 989–993.
Sackstone WE (1979) Treatment with Ridomil to control downy mildew of sunflower. Sunflower Newsl 3:7.
Saha T, Sukul P (1997) Metalaxyl—its persistence and metabolism in soil. Toxicol Environ Chem 58:251–258.
Salau JS, Alonso R, Batllo G, Barcelo D (1994) Application of solid phase disk extraction followed by gas and liquid chromatography for the simultaneous determination of the fungicides: captan, captafol, carbendazim, chlorothalonil, ethirimol, folpet, metalaxyl and vinclozolin in environmental waters. Anal Chem Acta 293:109–117.
Scarponi L, Martinetti L (1992) Investigation on the presence of residues of metalaxyl and penconazole in Italian wine. Vignevini 19:59–62.
Schwinn F (1983) In: Erwin DC, Bartnicki-Garcia S, Tsao PH (eds) *Phytophthora:* its biology, taxonomy, ecology. American Phytopathology Society, St. Paul, MN, p 327.
Schwinn F, Staub T, Urech PA (1977) A new type of fungicide against diseases caused by Oomycetes. Meded Fac Landbouwwet Rijksuniv Gent 42:1181–1188.
Seefeld F, Dunsing M (1990) Gas chromatographic determination of metalaxyl in plant materials. Z Gesamte Hyg Inre Grenzgeb 35:535–537.
Sethunathan N, Siddaramappa R (1978) Microbial degradation of pesticides in rice soils. In: Soils and Rices. International Rice Research Institute, Los Banos, Philippines, pp 479–497.
Sharom MS, Edgington LV (1982) The adsorption, mobility and persistence of metalaxyl in soil and aqueous system. Can J Plant Pathol 4:334–340.
Sharma D, Awasthi MD (1997) Adsorption and movement of metalaxyl in soils under unsaturated flow conditions. Plant Soil 195:293–298.
Sharma KD, Kolte SJ (1985) Metalaxyl in the control of downy mildew and white rust of rapeseed and mustard. Pestology 9:31–35.
Shetty PK, Magu SP (1997) Influence of metalaxl on *Glomus fasciculatum* associated with wheat (*Triticum aestivum* L.). Curr Sci 72:275–277.
Siddiqui MR, Siddiqui KA, Gour A (1987) Economic control schedule for downy mildew of bazra hybrids. Seed Res 15:201–205.
Singh A, Tewari AN, Rai RC (1985) Control of Karnal bunt of wheat by a spray of fungicides. Indian Phytopathol 38:104–108.

Singh US, Tripathi RK (1980) Estimation of the systemic fungicide ridomil by thin layer chromatography. J Chromatogr 200:317–323.
Singh US, Tripathi RK (1982) Physicochemical and biological properties of metalaxyl. Indian J Mycol Plant Pathol 12:287–294.
Singh US, Tripathi RK, Kumar J, Dwivedi TS (1986) Uptake, translocation, distribution and persistence of ^{14}C-metalaxyl in pearl millet (*Pennisetum americanum* (L.) Leeke). J Phytopathol 117:112–135.
Slawinski J, Puzyna W, Stawinska D (1978) Chemiluminescence during photooxidation of melanins as soil humic acids arising from a singlet oxygen mechanism. Phytochem Photobiol 28:459–463.
Smith CA, Iwata Y, Gunther FA (1978) Conversion and disappearance of methidathion on thin layers of dry soil. J Agric Food Chem 26:959–962.
Speck M, Dirr EJ (1980) Gas chromatograhic determination of metalaxyl (Ridomil) residues in tobacco. J Chromatogr 200:313–316.
Stone JR, Verma PR, Dueck J, Westcott ND (1987) Bioactivity of the fungicide metalaxyl in rape plants after seed treatment and soil drench applications. Can J Plant Pathol 9:260–264.
Sukop M, Cogger CG (1992) Adsorption of carbofuran, metalaxyl and simazine: K_{oc} evaluation and relation to soil transport. J Environ Sci Health B 27:565–590.
Sukul P, Moza PN, Hustert K, Kettrup A (1992) Photochemistry of metalaxyl. J Agric Food Chem 40:2488–2492.
Tafuri F, Marucchini C, Patumi M, Businelli M (1981) Gas chromatographic determination of metalaxyl in soils and sunflower. J Agric Food Chem 29:1296–1298.
Tomlin C (1995) A world compendium: the pesticide manual. Crop Protection Publications, British Crop Protection Council, UK, p 452.
Tonini A, Avigliano M (1981) Efficacy of metalaxyl against *Peronospora tabacina* and residues on tobacco. Ann Ist Sper Tab 8:63–68.
Towers GHN (1980) Photosensitizers from plants and their photodynamic action. In: Reinhold L (ed) Progress in Photochemistry. Oxford Press, pp 183–202.
Tripathi RK, Singh US (1983) Physico-chemical and biological property of metalaxyl. In: Hussain A, Singh K, Singh BP, Agnihotri VP (eds) Recent Advances in Plant Pathology. Print House (India), Lucknow, pp 201–226.
Urech PA, Eberle A, Ruess W (1978) Chemical control of downy mildews through soil application of ridomil. In: Abstracts, 3rd International Congress of Plant Pathology, Munich, August 16–23, 1978, p 363.
Urech PA, Schwinn F, Staub T (1977) CGA 48988, a novel fungicide for the control of late blight, downy mildew and related soil borne diseases. In: Proceedings of the British Crop Protection Conference, Brighton, Nov. 21–24, 1977, vol 2. Boots, Nottingham, p 623.
Van Bruggen AHC, Milgroom MG, Osmeloski JF, Fry WE, Jacobson JS (1987) Attenuation of metalaxyl on potato leaves by simulated acidic rain and residence time. Phytopathology 77:401–406.
Vasilieva GK, Galulin RV, Sukhoparova VP, Galulina RA, Bernat I, Shally A, Kaluz S, Ragala S (1991) Ecotoxicological evaluation of the fungicide ridomil in vineyards. Agrokhimiya 4:100–106.
Venugopal MN, Safeeulla KM (1978) Chemical control of downy mildew of pearl millet, sorghum and maize. Indian J Agric Sci 48:537–539.
Vuik J, Brouwer W, Krishnadath GJN, van de Lagemaat D (1989) Gas chromatographic determination of metalaxyl in lettuce. J Agric Food Chem 37:88–90.

Wang Hua Guo, Peng GY, Qi Meng Wen, Wang HG, Peng GY, Qi MW (1995) Study on degradation and residues of ^{14}C-metalaxyl in soil. Acta Agric Univ Pekin 21: 395–401.
Watkins DAM (1979) Photochemical studies on pesticides. Pestic Sci 10:181–182.
WHO (1991) Guide to codex maximum limit for pesticide residues, part 2. Codex Alimentarious Commission, FAO, Rome, Italy.
Williams RJ, Singh SD (1981) Control of pearl millet downy mildew by seed treatment with metalaxyl. Ann Appl Biol 97:263–268.
Wynn FC, Crute IR (1981) Uptake, translocation and degradation of metalaxyl in lettuce. In: Proceedings of the 1981 British Crop Protection Conference. British Crop Protection Council, London, pp 137–145.
Wynn FC, Crute IR (1983) Bioassay of metalaxyl in plant tissue. Ann Appl Biol 102: 117–121.
Yao JR, Liu SY, Freyer AJ, Minard RD, Bollag JM (1989) Photodecomposition of metalaxyl in an aqueous solution. J Agric Food Chem 37:1518–1523.
Zaki AI, Zentmeyer GA, LeBaron HM (1981) Systemic translocation of ^{14}C-labelled metalaxyl in tomato, avocado and *Persea indica*. Phytopathology 71:509–514.
Zheng Z, Shu-Yen L, Freyer AJ, Bollag JM (1989) Transformation of metalaxyl by the fungus *Sycephalastrum racemosum*. Appl Environ Microbiol 55:66–71.

Manuscript received April 21, 1999; accepted May 29, 1999.

Indoor Household Pesticides: Hazardous Waste Concern or Not?

John M. Owens · Patrick D. Guiney · Philip H. Howard · Dallas B. Aronson · D. Anthony Gray

Contents

I. Introduction	27
II. Definition of Indoor Household Pesticides	28
III. Options for Household Hazardous Waste Collection and Disposal	30
IV. Potential Risk from Disposal of Indoor Household Pesticides into Landfills	38
A. Leaching Potential of Insecticides, Repellents/Attractants, and Rodenticides	38
B. Leaching Potential of Disinfectants and Sanitizers (Antimicrobials)	43
C. Amount of Household Pesticides Disposed into Solid Waste Landfills	45
D. Assessment of Human Health Risk	46
E. Improved Standards for Solid Waste Landfills	57
Summary	60
References	61

I. Introduction

The federal Resource Conservation and Recovery Act (RCRA) sets basic standards for municipal and industrial waste disposal in the U.S., and charges the United States Environmental Protection Agency (EPA) with establishing and enforcing regulations that guide waste disposal practices of governmental and private entities. RCRA specifically defines all solid wastes generated by households as not hazardous and exempts that waste from the hazardous waste disposal requirements that are placed primarily on a wide variety of industrial wastes under Subtitle C. However, despite these statutory distinctions, the EPA has recommended that state and local waste disposal officials consider a number

---

Communicated by George W. Ware

J.M. Owens (✉)
S.C. Johnson & Son, Inc., 1525 Howe Street, M.S. #149, Racine, WI 53403-2236.
E-mail: jmowens@scj.com

P.D. Guiney
S.C. Johnson & Son, Inc., Racine, WI 53403, U.S.A.

P.H. Howard·D.B. Aronson·D.A. Gray
Syracuse Research Corporation, North Syracuse, NY 13212, U.S.A.

of categories of household waste (Table 1), including household pesticides[1] and certain other chemical specialty products, for segregation from the normal municipal solid waste (MSW) stream and special management as household hazardous waste (HHW) (USEPA 1986). After collection and segregation, this waste can either be reused (e.g., quantities of compatible leftover paints can be swapped, or combined and used), recycled (e.g., used motor oil and lead-acid batteries), treated (e.g., neutralization of acids or bases) and disposed as nonhazardous waste, or disposed with other Subtitle C hazardous waste in designated hazardous waste landfills or incinerators. HHW program personnel also dispose of some wastes brought to their programs directly to Subtitle D facilities or municipal sewer systems, but this disposal occurs after the expense of segregated collection and sorting of HHW has been sustained.

While costs can vary, disposal of HHW in the more highly developed programs typically costs about 15 times as much as normal MSW disposal (e.g., in California $123/t for MSW, versus $1943/t for HHW according to a Tellus Institute Report citing 1990 data [Tellus Institute 1991]). In 1986, the EPA reported that HHW collection costs in three cities ranged from $1.24 to $9.05/lb, the latter figure including waste oil collection (USEPA 1986). Although in more recent times some HHW programs around the U.S. have further reduced management costs for the wastes they collect (Waste Watch Center 1993), this continues to be a very costly disposal option compared to collection and disposal of normal trash (MSW). Regional HHW programs report that the average participant brings 50 to 100 lb of HHW to these collection programs, with much of the weight in leftover paint, at a cost to the program of about $1.00/lb or somewhat higher (USEPA 1986). Thus, it is appropriate for all parties involved in addressing the HHW issue to carefully and objectively examine, based on scientific criteria and reasonable judgment, the true need to involve each of the various categories of household products or other materials in these costly programs.

II. Definition of Indoor Household Pesticides

As regulated by the EPA under the Federal Insecticide, Fungicide and Rodenticide Act (FIFRA), "pesticides" include all chemical active ingredients and formulated products, organisms (beneficial microorganisms), or devices that are marketed to kill, repel, or otherwise mitigate a wide variety of pests. Conventional pesticidal chemicals are categorized by terms such as insecticides, herbicides, fungicides, bactericides, or rodenticides, etc., based on their target pests. This review focuses on household pesticides that are specifically formulated as ready-to-use products for sale to and use by homeowners and institutional users (e.g., janitorial personnel) to control pests indoors. These are primarily consumer insecticides, mosquito repellents, disinfectants and sanitizers (antimicro-

---

[1]Common and technical names for indoor household pesticides listed in this article are given in Table 11, on page 58.

Table 1. Examples of household hazardous waste categorizations.

Rathje et al. (1987)	U.S.EPA (1986)	Bomberger et al. (1988)	Bertrand et al. (1995)	Sack et al. (1992)
Household cleaners	Household cleaners	Household cleaners	Nonaerosol cleaners	Household cleaners
Automotive maintenance	Automotive products		Automobile-related materials	Automotive products
	Household maintenance	Household maintenance		
		Household polishes		Household polishes
Pesticides and yard maintenance	Lawn and garden products	Pesticides and yard maintenance	Pesticides	
Batteries and electrical		Batteries	Batteries	
Prescription drugs				
Selected cosmetics				Cleaners for electronic equipment
				Oils, greases, lubricants
				Adhesive-related products
				Fabric and leather treatments
			Paints	Paint-related products
			Acids, bases, oxidants	
Miscellaneous	Miscellaneous		Miscellaneous	Miscellaneous

bials), and other products, including pet care products, home-use rodenticides, and specialty medical/veterinary products such as lice shampoos; all of which are used primarily to control pests of recognized public health significance. The latest EPA Pesticide Industry Sales and Usage report estimates that of 100 million U.S. households in 1997, 56 million used insecticides, 17 million used repellents, and 42 million used disinfectants (USEPA 1998e). Similar ready-to-use insecticides and the many forms of antimicrobial products (ready-to-use or concentrates that are diluted, usually with water, before use) that are used indoors by institutional users are also defined here as indoor household pesticides.

Table 2 lists the active ingredients in various categories of indoor household insecticide, repellent/attractant, and rodenticide products. The insecticides were identified in a detailed market survey of indoor household and pet care pesticide products conducted in 1996 by SC Johnson. In that study, samples of all products from 22 subcategories available at 10 retail outlets within each of 10 market areas across the U.S. (selected by highest sales volume for these product categories) were purchased. Approximately 400 separate household insecticide products were collected. The list of rodenticide active ingredients was obtained from animal damage control experts at Purdue University's Center for Urban and Industrial Pest Management.

Indoor household pesticides, as considered here, do not include products that are considered "restricted use" by the EPA (USEPA 1992d) and that can only be sold to and used by specially trained and certified applicators, but do include only those that are considered "general use." Paints containing mildewicides or insecticides, wood preservatives, and lawn and garden pesticides are not considered indoor household pesticides here; these are generally considered as distinct waste categories by HHW programs (Sack et al. 1992; see Table 1), with the waste management status often depending on the type of product and its formulation and condition. Lawn and garden pesticides, which include insecticides, herbicides, fungicides, and animal repellents, are excluded because they are not used indoors and are commonly sold as concentrate formulations that require dilution before application (disposal would be primarily of the concentrate).

III. Options for Household Hazardous Waste Collection and Disposal

The primary environmental protection objective of HHW programs is to avoid negative environmental impacts from disposal of products designated as HHW, as would otherwise occur by disposal into the normal MSW stream or by inappropriate disposal in backyards, down household drains, into storm sewers, or elsewhere. This objective is in contrast to other product safety concerns such as exposures during normal usage or accidental situations routinely addressed by state and federal regulatory agencies (e.g., EPA, U.S. Consumer Product Safety Commission (CPSC), U.S. Food & Drug Administration (FDA), U.S. Occupational Safety and Health Administration (OSHA), U.S. Department of Transportation (DOT)) under other statutes (Rathje 1998). Local governments in a number of states have addressed HHW disposal, including disposal of household

Table 2. Common active ingredients in indoor household insecticide, repellent/attractant and rodenticide products.

Product type and active ingredient	Typical uses	Water solubility[a] (ppm)	Log K_{ow}^a	K_{oc}^a (mL/g)	Hydrolysis $t_{1/2}^a$ (d)	Soil $t_{1/2}^a$ (d)
Organophosphate insecticides:						
Chlorpyrifos (2921-88-2)	Residual[yy]	1.2 0.71[b]	5.27[c]	9,930 17,000[b]	77 (pH 5); 29 (pH 7); 16 (pH 9) 73[b]	30 (aer) 88(aer)[b] 140(an)[b]
Diazinon (333-41-5)	Residual	60	3.30	1520 1200[b]	12 (pH 5); 138 (pH 7); 77 (pH 9) 20[b]	39(aer) 17(aer)[b] 35(an)[b]
Dichlorvos (62-73-7)	Nonresidual	8,000	2.29[d]	50	31.5 (pH 4); 3 (pH 7); 2 (pH 9)	3.5(aer)[e] 3.5(an)[e] 1.5–17[f,g]
Trichlorfon (52-68-6)		130,000	0.43	15	103 (pH 5); 1.4 (pH 7); 0.02 (pH 9)	6.4 (aer) 1.8 (an)
Tetrachlorvinphos (22248-79-9)	Pet care	11	3.53[c]	1360[h]	54 (pH 3); 44 (pH 7); 3.3 (pH 10.5) at 50 °C[i]	15–22 (aer)[j]
Naled (300-76-5)	Pet care	1.5	1.38[c]	157	14[k]	4 (aer)

[a]Data obtained from the USDA Soil Conservation Service/Agriculture Research Service (SCS/ARS) Pesticide Properties Database (1997) (Wauchope et al. 1991), unless otherwise indicated.
[b]Johnson 1991, [c]Hansch et al. 1995, [d]Hansch and Leo 1981, [e]Kawamoto and Urano 1990, [f]Korotova and Denchenko 1978.
[g]Menzie 1972, [h]Leistra et al. 1984, [i]Tomlin 1997, [j]Domsch 1984.

Table 2. (Continued).

Product type and active ingredient	Typical uses	Water solubility[a] (ppm)	Log K_{ow}^a	K_{oc}^a (mL/g)	Hydrolysis $t_{1/2}^a$ (d)	Soil $t_{1/2}^a$ (d)
Carbamate insecticides:						
Propoxur (114-26-1)	Residual	1,800	1.52[c]	29	99 (pH 7); 1 (pH 9)	145(aer)/94(an)[i] 103(aer)[i] 115(an)[l]
Bendiocarb (22781-23-3)	Residual	260	1.70	570	5 (pH 5); 3.5 (pH 7); 0.03 (pH 9)	3.5 (aer) 4–6[m]
Carbaryl (63-25-2)	Pet care	110	2.31	288 360[b]	1.9–2,100 (pH 4.5–9, 25 °C)[n] 10.5 (pH 7)	17 (aer)/46 (an) 8(aer)[b] 76(an)[b]
Methomyl (16752-77-5)	Fly strip/trap	57,900	0.093	86	stable (pH 5,7) 35 (pH 9)	30 (aer)
Pyrethroid insecticides:						
Pyrethrin (8003-34-7)	Nonresidual	Insoluble	6.15[o]	10,000[p]		
Pyrethrin I (121-21-1)	Nonresidual	Insoluble	6.28[o]	10,000[p]	Stable (pH 5 & 7); 17 (pH 9)[q]	3.2(aer)/86 (an)[q]
Pyrethrin II (121-29-9)	Nonresidual	Insoluble	5.33[o]	3,030[p]		
Piperonyl butoxide (51-03-6)	Synergist	~1,000[n]	4.75[c]	70[p]	Stable (pH 5, 7, 9)[q]	49(aer)/144(an)[q]

[k] Gustafson 1989.
[l] Kanazawa 1987, [m] Adcock et al. 1975, [n] Chapman and Cole 1982, [o] Meylan and Howard 1995.
[p] Meylan et al. 1992, [q] Technical Data given to S.C. Johnson Wax by Supplier 1995–1996.

Table 2. (Continued).

Product type and active ingredient	Typical uses	Water solubility[a] (ppm)	Log K_{ow}^a	K_{oc}^a (mL/g)	Hydrolysis $t_{1/2}^a$ (d)	Soil $t_{1/2}^a$ (d)
MGK-264 (113-48-4)	Synergist	insoluble[i]	3.70[i]	10,400[p]	Stable (pH 5, 7, 9)[i]	
Tetramethrin/Neopynamin (7696-12-0)	Nonresidual	1.8[i]	4.73[c]	3,500[p]		
Allethrin (584-79-2)	Nonresidual	insoluble[i]	4.78[c]	3,080[p]		
Sumithrin/Phenothrin (26002-80-2)	Nonresidual	<0.0097[i]	6.76[f]	180,000[p]	Stable, hydrolyzed by alkalis[i]	
Cypermethrin (52315-07-8)	Residual	0.004	6.6	61,000	Stable (pH 5, 7) 1.8 (pH 9)	36 (aer)/36 (an)
Deltamethrin (52918-63-5)	Residual	<0.002[i]	6.20[c]	460,000–1,630,000[i]	Stable (pH 5 & 7); 2.5 (pH 9)[i]	35–56 (aer)[s] 11–19[q]
Resmethrin (10453-86-8)	Short residual	<1 (30 °C)[f]	6.14[u]	425,000[c]	89 (pH 5); 168 (pH 7); 127 (pH 9)[q]	197.5 (aer)/682 (an)[q]
Esbiothrin (28434-00-6)	Short residual	4.6[q]	4.68[c]	1134–1718[q]	1411 (pH 5); 547 (pH 7); 4.3 (pH 9)[q]	16.9–22 (aer)[q]
Permethrin (52645-53-1)	Residual	0.006–0.2	6.1	39,300	Stable (pH 5, 7) 50 (pH 9)	30 (aer) 108(an) 32–34 (an)[v]
Fenvalerate (51630-58-1)	Residual	<0.002	6.2[w]	5,273	15–90, in soil[w] 4–15, in water[w] Stable (pH 5–9)	163(aer) 15–90(aer)[x] 190(an)[x]

[f]Sumitomo Chemical Co. Inc, 1992, [s]Hill and Schaalje 1985, [i]Shiu et al. 1990.
[u]Noble 1993, [v]Jordan and Kaufman 1986, [w]IARC 1991, [x]Ohkawa et al. 1978.

Table 2. (Continued).

Product type and active ingredient	Typical uses	Water solubility[a] (ppm)	Log K_{ow}^a	K_{oc}^a (mL/g)	Hydrolysis $t_{1/2}^a$ (d)	Soil $t_{1/2}^a$ (d)
Esfenvalerate (66230-04-4)	Residual	<0.002	>4.0	5273	Stable (pH 5, 7, 9)	108 (aer)
Cyfluthrin (68359-37-5)	Residual	0.02	5.95	31,000	Stable (pH 5)	4–90
Tralomethrin (66841-25-6)	Residual	0.08[i]	5[i]	43,796–675,667[i]	230 (pH 7); 2 (pH 9)	60 (aer) 64–84[i]
Other insecticides:						
Boric acid (10043-35-3)	Baits, dust	50,000[t]	−0.757[y]	11[z]		
Hydramethylnon (67485-29-4)	Baits	0.006	2.31	730,000		18 (aer)
Sulfluramid (4151-50-2)	Baits	Insoluble[j]	8.54[o]	3,500,000[p]		
Abamectin (71751-41-2)	Baits	0.01[aa]	3.99[q]	5,000	>28 (pH 5, 7, & 9)[q]	14–60 (aer)[bb] >737 (an)[q]
Fenoxycarb (72490-01-8) ("use in decline")	Growth regulator	5.7	4.07	3,220	Stable (pH 5, 7, 9)	51–75(aer) 83–230(an)
Pyriproxyfen (95737-68-1)	Growth regulator		5.55[o]	405,000[p]		
Methoprene (40596-69-8)	Growth regulator	1.4[t]	5.50[c]	2830[p]	Stable[j]	<10 (aer)[cc] 10–14 (an)[dd]
Hydroprene (41096-46-2)	Growth regulator	0.54[t]	6.73[o]	12,000[p]		

[y]U.S Barox MSDS 1993, [z]Lyman et al. 1990, [aa]Budavari et al. 1996.
[bb]USEPA 1990b, [cc]Zoecon Technical Bulletin 1989, [dd]USEPA 1991c.

Table 2. (Continued).

Product type and active ingredient	Typical uses	Water solubility[a] (ppm)	Log K_{ow}^a	K_{oc}^a (mL/g)	Hydrolysis $t_{1/2}^a$ (d)	Soil $t_{1/2}^a$ (d)
Limonene (138-86-3)	Nonresidual	8.7[ee]	4.57[ff]	1,300[p]		2–10 (aer)[gg]
Linalool (78-70-6)	Nonresidual	1,590[ee]	2.97[ff]	56[p]		122–128 (aer)[hh]
Fipronil (120068-37-3)	Baits, Pet	2.3[hh]	4.01[hh]	700–4300[ii]	Stable (pH 5, 7) 28 days (pH 9)[hh]	116–130 (an)[hh]
Petroleum oil	Cave					
Repellents and attractants:						
DEET (134-62-3)	Skin application	>1,000[jj]	2.34[kk]	540[p]		Slow (aer)[ll] >330 (an)[mm]
Z-9 Tricosene (27519-02-4)	Pheromone	0.3[i]	4.08[i]	4,900,000[p]		
Citronella oil (8000-29-1)	Skin and candles		3.53[o]	148[p]		
Pennyroyal (8007-44-1)						
MGK 326 (136-45-8)	Sprays	Insoluble[nn]	3.57[nn]	420[p]		
Pine (8002-09-3) and patchouli oils (8014-09-3)						
Moth proofers:						
Naphthalene (91-20-3)		31[oo]	3.30[oo]	871[pp]		2.1, 2.2 (aer)[qq] Few days to >80 (aer)[rr] 40–50[ss]

[ee]Yalkowsky and Dannenfelser 1992, [ff]Li and Perdue 1995, [gg]Misra et al. 1996, [hh]USEPA 1996b, [ii]Bobe et al. 1997. [jj]Chemicals Inspection and Testing Institute 1991, [kk]Suryanarayana 1991, [ll]Rogers et al. 1986, [mm]Kuhn and Suflita 1989. [nn]Tomlin 1994, [oo]Howard and Meylan 1997, [pp]Howard 1989, [qq]Park et al. 1990, [rr]Grenney et al. 1987.

Table 2. (Continued).

Product type and active ingredient	Typical uses	Water solubility[a] (ppm)	Log K_{ow}^a	K_{oc}^a (mL/g)	Hydrolysis $t_{1/2}^a$ (d)	Soil $t_{1/2}^a$ (d)
p-Dichlorobenzene (106-46-7)		76[oo]	3.44[oo]	700[pp]		14 (aer)[tt] 385 (an)[uu]
Lavandin oil (8022-15-9)						
Cedarwood oil (8000-27-9)			1.28[o]	78[z]		
Rodenticides:						
Warfarin (81-81-2)		17	2.70[c]	919	5,900 (pH 7, 25 °C)[vv]	
Brodifacoum (56073-10-0)		<10[t]	8.51[o]	7,500,000[p]	Stable (pH 4, 7, & 9)[ww]	56–84 (aer)[ww]
Cholecalciferol (67-97-0)		Insoluble[nn]	10.24[o]	1,500,000[p]		
Diphacinone (82-66-6)		0.3[i]	4.85[o]	74,000[p]	14 (pH 6–9); <1 (pH 4)[j]	
Bromadiolone (28772-56-7)		19[t]	7.02[o]	210,000[p]		
Chlorophacinone (3691-35-8)		100[j]	5.50[o]	15,556–135,976[j]	Stable[i]	
Pindone (83-26-1)		18[ee]	2.87[o]	197[p]		
Zinc phosphide (1314-84-7)		Insoluble[i]			2 (pH 4); 14.5 (pH 7); 239 (pH 7.9)[xx]	

[ss]Al-Bashir et al. 1990, [tt]Bjerg et al. 1996, [uu]Masunaga et al. 1996, [vv]Ellington et al. 1988, [ww]USEPA 1979.
[xx]Hilton and Robison 1972, [yy]"Residual" insecticides are applied to surfaces leaving a deposit or residue that controls insects coming into contact with the residue.

pesticide products, within the context of their overall MSW programs either by use of intermittent "clean sweep" days or establishment of routine, ongoing programs that incorporate use of permanent facilities, some augmented by mobile collection units. Although some of these HHW programs have focused on segregating only used motor oil, antifreeze, paints, and other architectural coatings (especially oil-based), and batteries (especially those containing lead, nickel, cadmium, or mercury); other programs have included a variety of other household product categories such as solvents or paint strippers, oven and drain cleaners, spot removers, adhesives, pesticides (of all types, including disinfectants and sanitizers), solvent-based waxes or polishes (furniture, auto, shoe, floor), and other household cleaning products. Many states also maintain special collection programs for agricultural pesticides and pesticide containers; these programs are targeted at banned or unusable pesticides (Cubbage 1992). The numbers of these HHW programs have grown significantly in recent years and are expected to continue to do so over the next several years (Waste Watch Center 1997). This trend is expected to increase dramatically if proposed municipal liability changes to the federal Superfund law (CERCLA) are ultimately adopted. Amendments to CERCLA have been proposed that would limit municipal liability for cleanup costs at individual Superfund sites for municipalities that conduct HHW collection programs.

Indoor household pesticide products not collected separately as HHW remain in the normal stream of MSW and are disposed of most commonly by landfilling, incineration, or through other waste diversion programs such as reuse, recycling, and composting. Landfilling is the most common waste management method, with 2,514 landfills operating in the U.S. as of 1997 (Glenn 1998); approximately 61% of all MSW is landfilled (Glenn 1998). Incineration is also a common waste management method, with approximately 9% of all MSW incinerated (Glenn 1998). However, current policy trends clearly disfavor incineration among disposal options, even though properly functioning, modern MSW incineration methods will readily break down household pesticide residues in the MSW stream because their active ingredients are not thermally stable. In a study of 16 pesticides, including chlorpyrifos, which is a common indoor household pesticide, 99% decomposition was determined with a 2-sec retention time at temperatures of 700 °C (Tirey et al. 1993).

Municipal waste can also be diverted from the waste stream through recycling and composting programs. Nationwide, about 30% of MSW, including yard trimming composting, is recycled (Glenn 1998). Composting programs generally are preceded by source separation before disposal wherein product containers, including those of product wastes considered HHW, are expected to be removed and either recycled directly or disposed of in a landfill. If nonempty indoor household pesticide containers are inadvertently added to composting mixtures, studies by Lemmon and Pylypiw (1992), Vandervoort et al. (1997), and Michel et al. (1997) have shown that common household pesticides such as chlorpyrifos and diazinon are readily degraded during composting. Because landfilling is currently the predominant method of waste disposal, with potential for environmental release of organic compounds, the focus of this review is on the possibility of groundwater contamination because of the inclusion of indoor household pesticides in the MSW stream with final disposal into a MSW landfill.

IV. Potential Risk from Disposal of Indoor Household Pesticides into Landfills

The desirability of separating indoor household pesticides for disposal into a hazardous waste disposal facility, as opposed to depositing them in a municipal sanitary waste landfill (as envisioned by the RCRA exemption from Subtitle C management for all household wastes), is assessed in the following sections. First, published criteria are used to determine whether a pesticide, based on its physical and chemical properties, can be expected to leach to groundwater. Second, the amount of indoor household pesticides actually found in MSW is presented. Third, the risk to human health is modeled assuming exposure via ingestion of contaminated groundwater. Finally, current design standards for municipal solid waste landfills are examined for their ability to protect the public from potential leaching of indoor household pesticides.

A. Leaching Potential of Insecticides, Repellents/Attractants, and Rodenticides

Physical and chemical properties of various pesticides and other organic chemicals that contribute to persistence and mobility through soil have received considerable study by industry, government, and academic researchers, and are well understood. Specific criteria used to identify pesticides that have the potential to leach into groundwater have been published by the California Department of Food and Agriculture (CDFA) (Johnson 1991; Miller et al. 1990; Wilkerson and Kim 1986). Similarly, the EPA's Office of Pesticide Programs (OPP) proposed regulatory criteria for determining those pesticides that will be classified as "restricted use", due in part to their potential to leach into groundwater (USEPA 1991a), after examining 25 of 45 pesticides that were both detected in groundwater and had complete persistence and mobility data. These criteria, along with those reported by the CDFA, are summarized in Table 3. In both cases, they were developed to predict the potential of a pesticide to leach through soil into groundwater but can also be useful for identifying the potential of a chemical to leach from a landfill. K_d is defined as the soil–water partition coefficient, K_{oc} is the soil adsorption index normalized to the organic carbon (oc) of the soil, and the soil half-life is given for reliable laboratory aerobic and anaerobic soil metabolism or field dissipation studies.

The physicochemical properties and environmental fate data of indoor household insecticides, repellents/attractants, and rodenticides (Table 2) were collected primarily from the USDA-SCS Pesticide Properties Database (Wauchope et al. 1991) unless otherwise noted. When compared with both the CDFA and the OPP criteria, many of the active ingredients used in indoor household insec-

Table 3. Previously proposed criteria identifying pesticides that have the potential to leach into groundwater.

Source	Water solubility (ppm)	K_d		K_{oc} (mL/g)		Hydrolysis or photolysis $t_{1/2}$ (d)		Soil $t_{1/2}$ (d)
USEPA (1991a)			≤5[a] OR	≤500[a]	AND	<10% degradation in 30 d	OR	>3 wk(aer)[b]
CDFA (Wilkerson and Kim, 1986)	>7 ppm		OR	<512	AND	>13 d	OR	>11 d
CDFA (Johnson 1991)	>3 ppm		OR	<1900	AND	>14 d	OR	>610 d(aer) >9 d(an)[c]

[a]OR, detected below 75 cm in the soil profile.
[b]Soil aerobic metabolism half-life.
[c]Soil anaerobic metabolism half-life.

ticides and rodenticides are not sufficiently persistent or are so strongly adsorbed that they are very unlikely to be found in the leachate of landfills. Table 4 lists those compounds that either exceed or potentially exceed both the mobility and persistence criteria published by the CDFA or the OPP, suggesting that they may be capable of leaching through soil under and around a MSW landfill. Bromadiolone and chlorophacinone, rodenticide baits that would be disposed as formulated food-based bait matrices, were not reported as leachers although, according to the CDFA criteria, they exceeded the water solubility value. Both

Table 4. Compounds from Table 2 that exceed or potentially exceed criteria set by the EPA and the CDFA.

Exceed EPA criteria	Potentially exceed EPA criteria	Exceed CDFA criteria	Potentially exceed CDFA criteria
Propoxur	Citronella oil[a]	Diazinon	Tetrachlorvinphos[b]
Methomyl	MGK 326[a]	Propoxur	Methomyl[b]
Piperonyl butoxide	Cedarwood oil[a]	Piperonyl butoxide	Esbiothrin[b]
Boric acid	Pindone[a]	Boric acid	Limonene[a]
Petroleum distillates		Fipronil	Linalool[b]
		Petroleum distillates	Citronella oil[a]
		DEET	MGK 326[a]
		Naphthalene	Cedarwood oil[a]
		p-Dichlorobenzene	Warfarin[a]
			Pindone[a]

[a]Compound exceeds mobility criteria; however, persistence data are not available.
[b]Compound exceeds mobility criteria but meets aerobic biodegradation/hydrolysis criteria; however, anaerobic biodegradation data are not available.

compounds were unusual in having very large reported K_{oc} and high water solubility values. Because the CDFA mobility criteria are more conservative, i.e., more likely to classify a compound as a leacher, than those reported by the OPP, mainly because of a much higher K_{oc} cutoff (<500 for OPP versus <1900 for the CDFA), more compounds were listed as having the potential to leach using the CDFA criteria. However, the CDFA's more stringent criteria are not necessarily more accurate. Using the CDFA criteria Johnson (1991) reported, that 14 of 27 nonleaching pesticides were misclassified as leachers, while only 4 of 23 leaching pesticides were misclassified as nonleachers.

Groundwater monitoring data show that several of the compounds listed in Table 4 as having leaching potential have been detected in groundwater. Table 5 lists data for the 17 pesticides that are common to Table 2 and the EPA's Pesticides in Groundwater Database: A Compilation of Monitoring Studies: 1971–1991, which contains monitoring data for 302 pesticides and related compounds in drinking water wells (public community, public noncommunity, and private) and nondrinking water wells (monitoring wells and others such as for irrigation or other industrial purposes) throughout the U.S. (USEPA 1992b). Despite their widespread agricultural use (Gianessi 1992; Tomlin 1997), these compounds were only infrequently detected in groundwater. Detection of those compounds, which are classified in this review as nonleachers due to strong soil adsorption values or short soil half-lives, may have occurred at site locations

Table 5. Data from EPA compilation of monitoring studies for pesticides in groundwater 1971–1991 for 17 pesticides used in indoor household products.

Insecticide	Total wells sampled	Wells with detections	Percentage of wells with detections	Wells with detections >MCL	Concentration range (µg/L)
Methomyl	23,250	154	0.66	0	Trace–20.00
Carbaryl	25,712	106	0.41	0	0.03–610
Chlorpyrifos	5,398	32	0.60	0	Trace–0.654
Diazinon	3,884	42	1.1	2	Trace–3.2
Trichlorfon	459	12	2.6	0	10.00
Fenvalerate	345	5	1.4	0	0.01–0.28
Propoxur	21,405	5	0.02	4	2.0–35.0
Permethrin	1,097	4	0.36	0	0.01–1.25
Cypermethrin	311	0		0	
Naled	247	0		0	
Dichlorvos	188	0		0	
Tralomethrin	188	0		0	
Tetrachlorvinphos	173	0		0	
Pyrethrins	144	0		0	
p-Dichlorobenzene	97	0		0	
Naphthalene	82	0		0	
Bendiocarb	4	0		0	

where the aquifer was close to the surface and the soil properties were conducive to leaching. These are conditions that would not be expected at a newly sited and operating MSW landfill.

Only seven of the compounds from Table 2 (i.e., *cis*- and *trans*-permethrin, diazinon [only qualitative], dichlorvos, carbaryl, propoxur, methomyl, and MGK 264) were included in the 5-yr National Survey of Pesticides in Drinking Water Wells (NPS) (USEPA 1990a; USEPA 1992a). None of these compounds was detected in drinking water. No positive detections were found for the 11 pesticides listed in Table 2 common to both lists (chlorpyrifos, 282 wells sampled; diazinon, 337; dichlorvos, 11; trichlorfon, 69; tetrachlorvinphos, 11) during a survey of California well water monitoring 192 pesticides (Miller et al. 1990). In the 1996 update to the California well water survey, no positive detections were found for the seven pesticides listed in Table 2 common to both lists (diazinon, 1674 wells sampled; dichlorvos, 11; tetrachlorvinphos, 12; propoxur, 104; carbaryl, 1442; methomyl, 798; permethrin, 11) (Bartkowiak et al. 1997). Naphthalene (1501 wells sampled) was reported in two unverified cases.

Both the leaching criteria and monitoring data reported here are for mainly agricultural soil environments where the organic content is fairly consistent from soil to soil and both aerobic and anaerobic degradation can occur. A landfill environment, in comparison, is characterized by acidic pH, high organic content, and primarily anaerobic biodegradation processes after an initial aerobic degradation phase of a few days once the waste is placed (Barlaz and Ham 1993). These factors may affect both the mobility and persistence of compounds disposed in a MSW landfill. A major review of organic contaminants found in MSW landfill leachate was published by the EPA in 1988 (USEPA 1988a). The pesticides listed in that summary include (range of ppb; average ppb; number of values) 2,4-D (7.4–220; 129; 7), 4,4-DDT (0.042–0.22; 0.1031; 16), endrin (0.04–50; 16.8; 3), lindane (0.017–0.023; 0.020; 2), and pentachlorophenol (3–470; 173; 3). Of these compounds 2,4-D is a known leacher, and lindane has borderline leaching properties based on mobility and persistence data. Pentachlorophenol, as the phenolate anion, has much greater mobility than the undissociated molecule. Under acidic conditions, as in most landfills, this compound is expected to be present in both dissociated and undissociated forms. The presence of endrin and 4,4-DDT in landfill leachate is problematic as neither compound has properties that would suggest the potential for leaching. These may represent sampling errors, as may occur if sampling was completed within the landfill where the pesticides may have been deposited (not evidence of leaching ability), or, possibly, these compounds may have been carried by cosolvents into the groundwater. Pesticides detected twice or less often in MSW landfill leachate were (times detected; times analyzed for) aldrin (0; 60), BHC-alpha, -beta, or -gamma (0; 62), chlordane (0; 62), chlorobenzilate (0; 19), dieldrin (0; 60), dinoseb (0; 19), disulfoton (0; 19), endosulfan (0; 60), famphur (0; 19), heptachlor (0; 62), heptachlor epoxide (0; 60), kepone (0; 19), methoxychlor (0; 31), pentachloronitrobenzene (0; 19), phorate (0; 21), Silvex or 2,4,5-TP (2; 14), and zinophos (0;19). As expected, most of these compounds (i.e., aldrin, BHC-alpha, -beta, or -gamma, chlordane, dieldrin, disulfoton, endosulfan, heptachlor, kepone, methoxychlor, and phorate) meet OPP criteria as nonleachers. Thus, the most comprehensive review of landfill leachate available, which covers 83 MSW landfills and 60 other landfills (i.e., about 1% of the 6000 landfills operating at the time), found little detection of pesticides, even in landfills sited and operating before initial regulation of these facilities under RCRA in 1980.

The EPA review of MSW leachate reported the presence of both naphthalene (range 2–202 ppm; average 32.4 ppm; 23 values) and p-dichlorobenzene (range 1–52.1 ppm; average 13.2 ppm; 12 values) in landfill leachate. These were both listed by the CDFA as having the potential to leach. Naphthalene was detected in 21 of 44 pre-1980 and 0 of 11 post-1980 landfill leachate samples, and p-dichlorobenzene was detected in 11 of 46 pre-1980 and 0 of 11 post-1980 landfill leachate samples. These compounds are both used as active ingredients in mothproofing products (i.e., mothballs, crystals, or flakes). p-Dichlorobenzene is also used in urinal blocks in institutional settings. These may be examples of compounds that require further examination, including toxicological effects, to determine whether they should be treated as hazardous waste with proper disposal through a hazardous waste program. This review considers this in a subsequent section.

Sufficient data are available to suggest that hydrocarbon mixtures or their components are expected to leach into groundwater. However, the quantity of hydrocarbons added to MSW landfills by indoor household pesticide disposal is expected to be considerably less than that added to the landfill from other sources such as automotive products, oil-based paints, paint removers, kerosene, lighter fluid, paint thinner, solvents, petroleum ether, gasoline, and varnishes.

Groundwater monitoring studies that monitored for piperonyl butoxide, pindone, MGK 326, boric acid, and most of the natural oils, compounds shown in Table 2 to have physicochemical and fate characteristics that may make them possibly susceptible to leaching, were not located. Boric acid and related compounds, such as borax, tetraborate, and sodium tetraborate decahydrate, are naturally found in the environment (Smith 1985). Mixtures such as citronella oil, pennyroyal, pine and patchoulli oils, lavandin oil, and cedarwood as well as the compounds limonene and linalool are derived from botanical sources (Bauer et al. 1988). These compounds are not expected to have any significant public health effect if released to the environment from a landfill.

In addition to the physicochemical and fate characteristics of household pesticides (Table 2) that limit their potential to leach into groundwater, the form in which the pesticide is marketed should also be considered. Therefore, the leaching potential of compounds classified as possible leachers (Table 4) in a landfill must be considered in context with their use within the household pesticide product category. If the compound is found bound to a fly resin strip or within a bait carrier (e.g., methomyl or fipronil), or to a plastic matrix, such as a pet collar (e.g., tetrachlorvinphos), only a very small percentage of the active ingredient will be available to leach into the surrounding environment.

B. Leaching Potential of Disinfectants and Sanitizers (Antimicrobials)

Table 6 lists active ingredients from disinfectants and sanitizers (Block 1991), or antimicrobials, used in household and institutional settings by category and individual species. Data on many of the physical and chemical properties relevant to environmental fate considerations, such as those available for most compounds listed in Table 2, are typically not available for these disinfectants and sanitizers, primarily because they are not intended for outdoor application. Thus, our discussion of the disinfectant and sanitizer category and the potential for its active ingredients to be released to the environment following disposal to MSW landfills is based on somewhat different considerations than were used for insecticides, repellents/attractants, and rodenticides.

The first three groups of active ingredients listed in Table 6, chlorine com-

Table 6. Categories and individual types (with CAS registry number) of active ingredients used in household and institutional disinfectants and sanitizers listed as pesticides by the EPA.

Chlorine compounds
    Sodium hypochlorite (7681-52-9)
    Calcium hypochlorite dihydrate (7778-54-3)
    Potassium hypochlorite (7778-66-7)
    Chlorine dioxide (10049-04-4)
Iodine and iodophors
    Iodine (7553-56-2)
    Polyvinylpyrrolidone-iodine (25655-41-8)
    Nonoxynol-9 iodine (11096-42-7)
Quaternary ammonium compounds
    N-Alkyl ($C_{12}/C_{14}/C_{16}$) dimethyl benzyl ammonium chloride (8001-54-5)
    N-Alkyl (C_{12}/C_{14}) dimethyl ethylbenzyl ammonium chloride (68956-79-6)
    Dialkyl (C_8/C_{10}) dimethyl ammonium chloride (7173-51-5)
Acid/anionic compounds
    Phosphoric acid/sodium dodecylbenzene sulfonate (7664-38-2/25155-30-0)
Peroxygen compounds
    Hydrogen peroxide (7722-84-1)
    Peracetic acid (79-21-0)
Phenolic compounds
    O-Phenylphenol/Dowicide 1 (90-43-7)
    O-Benzyl-p-chlorophenol (120-32-1)
    p-Tertiary amylphenol (80-46-6)
Pine oil (8002-09-3)
Aldehydes
    Formaldehyde (50-00-0)
    Gglutaraldehyde (111-30-8)
Alcohols
    Ethyl alcohol (64-17-5)
    Isopropyl alcohol (67-63-0)

pounds, iodine and iodophors, and peroxy compounds, are reactive agents that decompose readily under sanitary landfill conditions. For example, both the hypochlorites and iodine lose much of their bactericidal potency in the presence of organic matter (Gump 1979). Iodophors are antiseptics that act by releasing available iodine following dilution (Gump 1979).

In a study of household hazardous wastes, Bomberger et al. (1988) concluded that acids, alkalis, and oxidants are of little concern in MSW landfill leachates because the concentration of acids and alkalis in HHW is low compared with that generated by the deterioration of the waste, and the oxidants are expected to react with the refuse components. Thus, any reactive agents and acids or bases are of little concern when deposited in sanitary landfills.

Quaternary ammonium compounds are tightly bound to organic matter in soils and sediments (Boethling and Lynch 1992) and, therefore, would not leach from landfills. Anionic surfactants, such as alkylbenzene sulfonates, are fairly water soluble and, therefore, small amounts may leach from landfills. However, the amount released should be extremely small compared to the amounts released by way of septic systems and municipal water treatment plants (POTW), and these compounds have few health concerns (Cahn and Lynn 1983).

The simple aldehydes and alcohols used as disinfectants and sanitizers are also likely to be generated by biodegradation of the landfill waste, and the quantities added to the landfill from disposal of products with these ingredients are insignificant by comparison. In addition, these compounds are considered to be very biodegradable, at least under aerobic conditions (Gerike and Gode 1990) expected to prevail in MSW landfills during the first few days to weeks after the refuse is placed.

The cyclic terpene alcohols found in pine oil are produced by steam distillation of pitch-soaked pine wood and therefore are natural products commonly found in foods and the environment. For example, α-terpineol, a primary constituent in pine oil has been detected in pineapples (Binder and Flath 1989), apricots, and plums (Gomez et al. 1993), orange juice (Moshonas et al. 1994), and fried chicken (Tang et al. 1983). Thus, leachate from cyclic terpene alcohols should be of little environmental or public health concern.

The phenolic disinfectants have a long history of use because of their well-known antimicrobial activity and "long history of human exposure without any adverse effects" (Paulus and Genth 1983). Most phenolic household disinfectants are released to the environment through municipal sewage treatment facilities, and this exposure, at least for O-benzyl-p-chlorophenol, has been characterized as safe (Werner and Wilson 1982; Werner et al. 1983). In addition, river die-away studies of O-benzyl-p-chlorophenol and o-phenylphenol show that both compounds are readily degraded (Gonsior et al. 1984; USEPA 1995b). In comparison to exposure during household use of these compounds, exposure from phenolic disinfectants disposed in landfills should be insignificant.

In conclusion, antimicrobial compounds have a long history of safe use in household and hospital settings (USEPA 1992c). Human exposure is expected to be primarily from application of these products within the home or institutional setting. In comparison to their primary application, where these products are almost always used in cleaning operations that ultimately result in disposal through drains to on-site or publicly owned wastewater treatment systems, exposure from the small amounts that might leach from MSW landfills are expected to be of little concern.

C. Amount of Household Pesticides Disposed into Solid Waste Landfills

The very small quantities of indoor household pesticides actually disposed of into MSW landfills suggest that they should not be found contaminating groundwater from these modern disposal facilities. Detailed waste characterization studies at several MSW landfills since the early 1980s have shown that household pesticides constitute a very small portion of MSW. Items characterized as HHW compose about 0.1–0.2% (range, 0.00147% to <1.0%) of the MSW stream (Anonymous 1983; Kinman and Nutini 1988; Puget Sound Council of Governments 1985). In 1983, the Los Angeles County Sanitation District conducted waste characterization studies at two landfills (Anonymous 1983). MSW from households contained 0.0045% HHW, while mixed and commercial loads contained 0.28% HHW. Overall, 0.13% of the refuse mass examined was classified as HHW. At one landfill where the 2056 containers from HHW products were checked for contents, 92% were empty. The 2056 containers were placed into six categories as follows: household cleaning products (40%), automotive products (30%), personal products (16%), paint and related products (8%), insecticides and other pesticides (3%), and other products considered hazardous (4%). Therefore, it is estimated that only 0.006% of the total MSW stream was from the category "insecticides and other pesticides."

The Minnesota Pollution Control Agency, in a detailed solid waste characterization study, reported that 0.8% by weight (range, 0.5%–1.0% across five counties) of the MSW was hazardous waste (Fredrickson et al. 1992). All aerosol cans were considered as HHW regardless of whether they had contained food, toxic compounds, or other products, and 3259 containers from HHW products were separated into 10 categories: aerosols (64%), flammables/solvents (16%), flammable solids (0%), paints (13%), chlorinated products/other pesticides (2%), acids (1%), bases (1%), organic peroxides (0.01%), oxidizers (0.15%), and other products considered hazardous (~3%). Containers containing chlorinated products and other pesticides were estimated to compose only 0.016% of the total MSW stream; 25% of the HHW containers, 18.5% of the aerosol containers, and 54% of the discarded chlorinated products/other pesticides containers had contents remaining.

A HHW study conducted in Palm Beach County, FL, reported that approximately 0.1% of the generated MSW in the county is HHW (Bertrand et al. 1995). HHW was obtained from 117 residential MSW samples of about 200–300 lb each collected from refuse vehicles. Pesticide and nonaerosol cleanser/disinfectant categories composed 12% and 6.4%, respectively, of the total HHW collected. The category "pesticides" was further divided into fungicide, herbicide, insecticide, and rodenticide subcategories representing 0.2, 0.5, 104.6, and 0.0 t/yr (net weight basis), respectively. It was estimated that only 1.5% of the total insecticides present from residential use was actually diverted from the MSW stream to the hazardous waste collection facility. Waste characterization of MSW from family-generated sources alone identified 15 containers (6%) of nearly 250 items as containing indoor household pesticides. None of the 15 submitted indoor household pesticide containers was empty, with contents estimated at 10%–75% of the original product.

Rathje et al. (1987) conducted waste characterization studies in Marin County, CA, and New Orleans, LA. Types of HHW were similar in both areas, with 0.35%–0.40% of the total MSW found as HHW: 1.1% and 0.5% of the separated containers in New Orleans, and 1.9% and 1.7% of the separated containers in Marin County contained pesticide and pet maintenance products, respectively. Based on product remaining following disposal, determined by weight, 0.7% and 0.3% of all HHW collected in New Orleans and 6.1% and 1.4% of all HHW collected in Marin County were from pesticide and pet maintenance products, respectively.

D. Assessment of Human Health Risk

A fundamental component of any environmental decision making is the acknowledgment that the cost of eliminating all impacts of human activities is impossibly high and therefore regulatory decisions must often be based on incomplete scientific information (Ruckelshaus 1983). Our objective in this section is to provide a quantitative basis for comparing and prioritizing risks of human health effects of potentially leachable indoor household pesticides from a MSW landfill. To develop this risk assessment, an EPA model was used with composite data collected from landfills across the U.S. Exposure pathways and potentially exposed populations were determined and toxicological data were collected for the compounds of interest. These exposure and toxicological data were then integrated, giving a quantitative estimate of risk for the conditions studied (USEPA 1989).

Only a small percentage of the total volume of MSW has been shown to be HHW and only a small percentage of that results from indoor household pesticides. In a landfill environment, with both dilution and dispersion processes occurring, very low concentrations of these pesticides are expected in either landfill leachate or groundwater. Very few published concentrations of the compounds listed in Table 2 inside a landfill or in leachate are available. Average concentrations and concentration ranges for naphthalene and p-dichlorobenzene in landfill leachate were reported earlier (USEPA 1988a). Two quantitative studies of household chemical products disposed in landfill facilities reported mean concentrations of 29, 18, and 43 mg/kg municipal solid waste for light-duty liquid detergents; 13, 0, and 33 mg/kg for toilet bowl cleaners; and 18, 5, and 5 mg/kg for bleach for sites in King County, WA; New Orleans, LA; and Marin County, CA, respectively (Savage and Miller 1997). An insecticide concentration of 155 ppm was estimated for MSW from a Palm Beach, FL study, based on disposal of 104.6 t of insecticides/yr (net weight basis) (Bertrand et al. 1995). These values are well within the concentration range of 0.05–10,000 ppm used here to model the environmental fate of indoor household pesticides in groundwater.

The principal route of exposure to indoor household pesticides from MSW landfills was assumed to be leaching through subsurface soils to groundwater and the subsequent ingestion of contaminated groundwater by a downgradient population. The impact of indoor household pesticide disposal into landfills was modeled for those compounds in Table 2 that had published EPA Reference Dose values for oral intake (RfD) and/or oral cancer slope factor (Q_1^*) values (Table 7) using the EPA's composite model for leachate migration with transformation products (EPACMTP) (Kool et al. 1994; USEPA 1993a; USEPA 1994b). This model was chosen for a number of reasons, including its state-of-the-art design and Monte Carlo capabilities, but especially because its use is consistent with EPA methodology. EPACMTP has been used by the EPA in their proposed rule for the identification and listing of hazardous waste (USEPA 1995a), which also presents a description of the model.

Briefly, EPACMTP is a numerical groundwater model that calculates the concentration of a contaminant from its source at the bottom of a landfill or other waste unit down through the unsaturated zone to the saturated zone and then to a downgradient drinking water well placed in the saturated zone. A number of assumptions are made by this model, including: (1) uniform contaminant flux across the bottom of the landfill (or other waste unit); (2) dilute concentrations of the contaminant are present; (3) steady-state and isothermal flows (unsaturated zone flow and transport is only downward, but saturated zone flow can be both advective and dispersive); and (4) uniform and porous soils and aquifer materials. Modeling options were chosen to be essentially consistent with EPA implementation for the hazardous waste identification and listing rule. Landfills were the only waste unit considered; surface impoundments, waste piles, and land application units were not considered because household wastes are not commonly disposed into these waste units. Distributions used during the Monte Carlo runs for such parameters as waste unit area, waste unit volume, and infiltration rate, as well as unsaturated and saturated zone parameters (e.g., aquifer pH, saturated thickness, porosity, temperature, and fraction of organic carbon), climatic data, and receptor well location were obtained from a national database compiled from an EPA Office of Solid Waste (OSW) survey of industrial Subtitle D waste management facilities for 790 landfills nationwide.

Fate and transport data used as input to the model were obtained from Table 2; a range of values was used for both soil degradation and soil adsorption parameters for most compounds. The main fate process in a landfill environment is soil degradation from biodegradation or possibly hydrolysis. The model accounts for only one combined first-order degradation rate, which is applied to both the unsaturated and saturated zones. Thus, if both hydrolysis and biodegradation occur, the individual rates are combined to a single overall rate. In practice this is only important when the two processes are similar in magnitude. However, to compare the effects of each fate process on downgradient concentrations of this set of compounds, hydrolysis and biodegradation were considered independently. The rate of hydrolysis is pH dependent; for those compounds with second-order rate constants (diazinon, carbaryl, warfarin, fenvalerate, esfenvalerate, fenoxycarb, hydroprene, MGK-326), pH distributions from the OSW survey were used during the Monte Carlo runs to model downgradient well concentrations. If only first-order rate constants were available, the model was run at a defined acid, neutral, and basic pH using an appropriate rate constant for each pH value. The model was also run several times for each compound using a range of biodegradation rates encompassing both aerobic and anaerobic conditions, when information was available. Anaerobic rate constant values were not available for tetrachlorvinphos, methomyl, tralomethrin, warfarin, MGK-326, bendiocarb, esfenvalerate, pyriproxyfen, and hydroprene. Only the parent compound was modeled; no daughter products were considered. Transport through the soil was represented by the measured or calculated K_{oc} value. Although sorption to soil material can be either linear or nonlinear (Freundlich), only linear sorption was considered.

EPACMTP can account for a contaminant source that is either constant with finite duration, constant with infinite duration, or decreasing with time (based on first-order decay). The first-order decay rate is based on the solution of the first-order differential equation describing the mass balance between the contaminant in the landfill and the mass being leached from the source. Results of simulations performed for this study were given for a finite (decaying) source with a 10,000 yr outlook. For each compound, the steady-state, peak, and 30-yr average concentrations, as well as the dilution attenuation factor (DAF), were reported by the model for a generic downgradient receptor well. The 30-yr average concentration is determined by identifying the peak concentration and then calculating the highest average concentration for a 30-yr period that contains the peak concentration. The DAF is simply the factor by which the initial concentration is reduced by the time it reaches the receptor well. Because each Monte Carlo run calculates a single DAF, ordering the DAFs in decreasing order provides the cumulative probability distribution. In keeping with EPA methodology, the 90th percentile was chosen. Because 2000 Monte Carlo iterations were performed for each chemical–K_{oc}–degradation rate combination, the 90th percentile was identified by simply sorting the DAF values and choosing the 1800th run. The peak receptor well concentration was used in the calculation of hazard quotient values, while the 30-yr average receptor well concentration was used in the calculation of lifetime cancer risk for those compounds in Table 2 with RfD or Q_1^* values (USEPA 1995a).

Available acute and chronic toxicity data, oral RfD, Q_1^* values, and carcinogenicity classifications (see Table 7) were collected for pesticides in Table 2 as a basis for determining human health risks of these compounds caused by ingestion of contaminated groundwater. Category definitions used for the carcinogenicity data are provided in Table 8. The oral RfD value is used to evaluate the noncancer health effect of a compound from chronic oral exposure; it provides an estimate of the daily exposure to a specific compound in mg/kg/d that will not result in an "appreciable risk of harmful noncancerous effects during the lifetime of a human" (USEPA 1989). Q_1^* values are used to estimate the carcinogenic potential of a compound over a human lifetime. EPA classified Group D and E compounds are not considered to be carcinogenic (USEPA 1995a).

Both noncancer health effects, measured as the hazard quotient, and cancer risk were calculated using the following equations as recommended by EPA (USEPA 1989) where:

$$HQ = \frac{(C_w)(IR_w)(EF)(ED)}{(BW)(AT)} \div RfD$$

$$Risk = \frac{(C_w)(IR_w)(EF)(ED)}{(BW)(AT)} Q_1^*$$

where:

HQ	=	hazard quotient
Risk	=	upper-bound lifetime cancer risk
C_w	=	contaminant concentration in drinking water
IR_w	=	drinking water ingestion rate (L/d)
EF	=	exposure frequency (d/yr)
ED	=	exposure duration (yr)
BW	=	body weight of an exposed individual (kg)
AT	=	averaging time (ED × 365 d for noncarcinogens; 25,550 d for carcinogens)
RfD	=	oral reference dose (mg/kg/d)
Q_1^*	=	oral potency slope [(mg/kg/d)$^{-1}$]

Published values for ingestion rate (2 L of water/d), exposure frequency (350 d/yr), exposure duration (30 yr), and body weight (70 kg) were used in these equations (USEPA 1997a). These values, except for body weight, are based on the 90th percentile of measured values and thus the risk calculations should present a conservative estimation of risk. Hazard quotients, values used to measure noncancer health effects, were calculated for an adult exposure with a hazard quotient of less than 1 defined as a "safe" exposure (USEPA 1990c).

Table 9 reports the estimated hazard quotient values for 28 indoor household pesticides. Downgradient receptor well concentrations were determined considering biodegradation and hydrolysis as separate fate processes; in nearly all cases, biodegradation alone is predicted to result in a complete or nearly complete loss of the studied compounds during transport to the downgradient well. For ease of comparison, if a hazard quotient was less than 0.01 due to biodegradation alone, a hazard quotient using the hydrolysis data was generally not presented. However, hazard quotient and cancer risk values based on biodegradation data estimated for tetrachlorvinphos, bendiocarb, methomyl, esfenvalerate,

Table 7. Toxicity of selected pesticides as technical materials.

Active ingredient	Oral LD$_{50}$[a] (mg/kg)	Dermal LD$_{50}$[b] (mg/kg)	Inhalation LC$_{50}$[a] (mg/m^3)	Chronic toxicity RfD (mg/kg/d)	Oral slope factor (Qf)[f] (mg/kg/d)	Carcinogenicity (weight of evidence classification)
Chlorpyrifos	82–165	2000	>200	0.003[c]		Not available (NTP[o]); Category E (USEPA[p])[m]
Diazinon	66–850	540–650	3500	0.0002 (MRL)[d]		Negative (NCI[q])
Dichlorvos	17–108	71–250	15	0.0005[c]	0.29[c]	Class 2B (IARC[r,s]; Category B2 (USEPA)[i]
Trichlorfon	438–560	2000 (rat)	1300			Group 3 (IARC)[t]
Tetrachlorvinphos	480	>2500	>854	0.03[c]		Group 3 (IARC)[t]
Naled	92–430	1100	>1500 (mouse)	0.002[c]		Category E (USEPA)[m]
Propoxur	70–191	>1000	>1440	0.004[c]	0.0037[m]	Category B2 (USEPA)[m]
Bendiocarb	34–156	>800	550	0.0013[e]		No effect in rats or mice[u]
Carbaryl	230–850	>2000	0.005–0.023 mg/kg	0.1[c]	0.0227[m]	Class 3 (IARC)[l]; Category C (USEPA)[m]
Methomyl	17–47	5880	77	0.025[c]		Unclassified
Pyrethrum	1300–2500	300, 5000	340			Deferred (USEPA)
Piperonyl butoxide	7500	1880	>5170			Class 3 (IARC)[l]; Category C (USEPA)[m]
MGK-264	2800–4990	470	>4.08			Category C (USEPA)[m]
Tetramethrin	>5000	>2000	2740			Category C (USEPA)[m]
Allethrin	585–1100	>2500	>2000			Unclassified
Sumithrin	>5000	>2000 (rat)	>1180, 3760			Unclassified
Cypermethrin	250–4150	>2500	2500	0.01[c]		Category C (USEPA)[m]
Deltamethrin	1335–5000	>2940	2200	0.01[f]		Class 3 (IARC)[s]
Esbiothrin	378–432	>2000	2630			Unclassified
Resmethrin	>2500	2500	>9490	0.03[c]		No effect in rats or mice[v]
Permethrin	>4000	>2000	23,500	0.05[c]	0.0184[m]	Class 3 (IARC)[s]; Category C (USEPA)[m]

Table 7. (Continued).

Active ingredient	Oral LD_{50}^a (mg/kg)	Dermal LD_{50}^b (mg/kg)	Inhalation LC_{50}^a (mg/m^3)	Chronic toxicity RfD (mg/kg/d)	Oral slope factor (Q_1^*) (mg/kg/d)	Carcinogenicity (weight of evidence classification)
Fenvalerate	70.2, 451	1000–3200	101	0.025^c		Class 3 (IARC)^s; Category E (USEPA)^m
Esfenvalerate	75–458	>2000	2930, 480–570	0.02^g		Class 3 (IARC)$^{s,\,aa}$; Category E (USEPA)^m
Cyfluthrin	900	>5000 (rat)	469	0.025^c		Category E (USEPA)^w
Tralomethrin	99–1250	>2000	>0.268	0.0075^c		Unclassified
Boric acid	3500–4100	>2000	28	0.09 (for boron and borates)^c		Negative (NTP); Category E (USEPA)^m
Hydramethylnon	1131	>5000	>5000	0.0003^c		Category C (USEPA)^m
Sulfluramid	543					Unclassified
Abamectin	10	>2000	1100	0.0004^h		Negative^x
Fenoxycarb	>16,800	>5000 (rat)	>480	0.8^i	0.056^i	Category B2 (USEPA)^m
Pyriproxyfen	>5000	>2000 (rat)	>1300	0.35^j		Category E (USEPA)^j
Methoprene	>34600	3038–10250	>210000			No effect in rats or mice^y
Hydroprene	>5100	>5100	>5000	0.1^k		Category D (USEPA)^m
Limonene	5500 ul/kg					No Evidence (DHHS/NTP)^z
Linalool	4180	>5000				Unclassified
Fipronil	97	354	390	0.0002^l		Category C (USEPA)^m
Petroleum oil	10000					Unclassified
DEET	1950	3180 ul/kg	>4100			Category D (USEPA)^m
Z-9 Tricosene	>23,070	>2025				Unclassified
Citronella oil	4380	>2000	>3.1			Unclassified
Pennyroyal oil						Unclassified
MGK-326	4270–5850	9500	>6.09		0.0012 (female) 0.0024 (male)^m	Category B2 (USEPA)^m

Table 7. (Continued).

Active ingredient	Oral LD_{50}^a (mg/kg)	Dermal LD_{50}^b (mg/kg)	Inhalation LC_{50}^a (mg/m^3)	Chronic toxicity RfD (mg/kg/d)	Oral slope factor (Q_f^*) (mg/kg/d)	Carcinogenicity (weight of evidence classification)
Pine oil	3200	5000				Unclassified
Patchoulli oil	>5000	>5000				Unclassified
Naphthalene	490–2600	>2000	>340	0.02^c		Category C (USEPA)^c
p-Dichlorobenzene	500–3863	>2000	>5.07	0.1 (MRL)^d	0.024^n	Group 2B (IARC)^f
Lavandin oil	>5000	>5000				Unclassified
Cedarwood oil	>5000	>5000				Unclassified
Warfarin	61–102	1400 (rat)	320	0.0003^c		Unclassified
Brodifacoum	0.27	200 (rat)	0.500–5.000			Unclassified
Cholecalciferol	42–619	>2000	138–380			Not available (NTP)
Diaphacinone	1.5	200 (rat)	2000			Unclassified
Bromadiolone	1.125	2.1	9230			Unclassified
Chlorophacione	20.5	200	3000			Unclassified
Pindone	10.3					Unclassified
Zinc phosphide	12	2000		0.0003^c		Not available (NTP)

^{a}Acute oral and inhalation tests were conducted in rats unless otherwise noted.
^{b}Acute dermal tests were conducted in rabbits unless otherwise noted.
^{c}USEPA 1991d, ^{d}ATSDR 1996, ^{e}USEPA 1994a, ^{f}USEPA 1998a, ^{g}USEPA 1998b, ^{h}USEPA 1996b.
^{i}USEPA 1997c, ^{j}USEPA 1998c, ^{k}USEPA 1997c, ^{l}USEPA 1997c, ^{m}USEPA 1996a, ^{m}USEPA 1996c, ^{n}USEPA 1997e.
^{o}NTP represents the National Toxicology Program, U.S. Department of Health and Human Services.
^{p}USEPA represents the US Environmental Protection Agency, ^{q}NCI represents the National Cancer Institute.
^{r}IARC represents the International Agency for Research on Cancer.
^{s}IARC 1991, ^{t}IARC 1987, ^{u}Hayes and Lawes 1990, ^{v}USEPA 1988b, ^{w}USEPA 1997a.
^{x}Lankas and Gordon 1989, ^{y}USEPA 1991c, ^{z}DHHS/NTP 1990.
^{aa}Weight of evidence classification for fenvalerate, a mixture of stereoisomers including esfenvalerate.

Table 8. Category definitions from agencies used to evaluate chemicals in Table 7.

United States Environmental Protection Agency (USEPA)
- Group A: Human carcinogen
- Group B: Probable human carcinogen
  - B1: Indicates limited human evidence
  - B2: Indicates sufficient evidence in animals and inadequate or no evidence in humans
- Group C: Possible human carcinogen
- Group D: Not classifiable as to human carcinogenicity
- Group E: Evidence of noncarcinogenicity for humans

International Agency for Research on Cancer (IARC)
- Group 1: The agent (mixture) is carcinogenic to humans
- Group 2A: The agent (mixture) is probably carcinogenic to humans
- Group 2B: The agent (mixture) is possibly carcinogenic to humans
- Group 3: The agent (mixture, exposure circumstance) is not classifiable as to its carcinogenicity
- Group 4: The agent (mixture, exposure circumstance) is probably not carcinogenic to humans

National Toxicology Program (NTP)/National Cancer Institute (NCI)
- Group 1: Known to be carcinogens where there is sufficient evidence of carcinogenicity from studies in humans that indicates a casual relationship between the agent and human cancer
- Group 2: Reasonably anticipated to be carcinogens
  - A. There is limited evidence of carcinogenicity from studies in humans, which indicate that casual interpretation is credible, but that alternative explanations, such as chance, bias, or confounding, could not adequately be excluded; or
  - B. There is sufficient evidence of carcinogenicity from studies in experimental animals which indicates that there is an increased incidence of malignant tumors: (a) in multiple species or strains, or (b) in multiple experiments (preferably with different routes of administration or using different dose levels), or (c) to an unusual degree with regard to incidence, site or type of tumor, or age at onset. Additional evidence may be provided by data concerning dose–response effects, as well as information on mutagenicity or chemical structure.

tralomethrin, hydramethylnon, pyriproxyfen, hydroprene, warfarin, and MGK-326 are qualified since actual anaerobic biodegradation data were not available for input into the EPACMTP model. Aside from a brief initial aerobic period when the waste is placed into the landfill, anaerobic conditions are expected to prevail in a landfill and groundwater plume environment. For this subset of compounds, hazard quotients are also presented based on degradation resulting from hydrolysis. When hydrolysis data are incorporated into the EPACMTP model, hazard quotients of <0.01 are reported under all modeled conditions of

Table 9. Estimation of hazard quotients from the ingestion of contaminated drinking water for selected compounds found in indoor household pesticides.

Compound	Hazard quotient	
	Biodegradation	Hydrolysis
Chlorpyrifos	<0.01	
Diazinon	<0.01	
Dichlorvos	<0.01	
Tetrachlorvinphos	**<0.01**[a]	<0.01
Naled	<0.01	
Propoxur	<0.01 to 0.3	<0.01–17
Bendiocarb	**<0.01**	<0.01
Carbaryl	<0.01	
Methomyl	**<0.01**	<0.01–11,200
Cypermethrin	<0.01	
Deltamethrin	<0.01	
Resmethrin	<0.01	
Permethrin	<0.01	
Fenvalerate	<0.01	
Esfenvalerate	**<0.01**	72–138,000
Cyfluthrin	<0.01	
Tralomethrin	**<0.01**	<0.01–209,000
Boric acid	Not biodegraded	606–31,000
Hydramethylnon	**<0.01**	<0.01
Abamectin	<0.01 to 0.02	
Fenoxycarb	<0.01	
Pyriproxyfen	**<0.01**	<0.01–7
Hydroprene	**<0.01**	0.14–25,900
Fipronil	<0.01	
Naphthalene	<0.01	
p-Dichlorobenzene	<0.01	
Warfarin	**<0.01**	4,530–600,000
Zinc phosphide	Not biodegraded	<0.01–13

[a]**Bolded** values are qualified as anaerobic biodegradation data were not available for use in these analyses, see text for additional discussion.

pH, concentration, and soil adsorption for tetrachlorvinphos, bendiocarb, and hydramethylnon, indicating that even when these compounds are not anaerobically biodegraded, hydrolysis alone would reduce concentrations to safe levels.

Although anaerobic biodegradation rate data were not located for methomyl, the reregistration eligibility decision (RED) for this compound states that methomyl is expected to biodegrade more rapidly under anaerobic conditions when compared to aerobic conditions (USEPA 1998d; Bromilow et al. 1986). Anaerobic data for fenvalerate indicate that esfenvalerate, one of four stereoisomers found in fenvalerate, is also likely to biodegrade under anaerobic conditions, resulting in a hazard quotient of 0 under the model conditions. Hydroprene and methoprene are structurally similar compounds; methoprene has a methoxy group on the alkyl chain that is not found in the hydroprene structure. An anaerobic half-life of 10–14 d in soil has been reported for methoprene (see Table 2), suggesting that degradation of hydroprene will be similarly rapid. Tralomethrin belongs to the pyrethroid group. Other pesticides within this group have been shown to biodegrade under anaerobic conditions (Table 2), with half-lives ranging from 36 to 682 d; the longest half-life was reported for resmethrin, which like tralomethrin has low soil mobility. Modeling of resmethrin using the 682-d biodegradation half-life resulted in a hazard quotient of 0. Based on this result, it is unlikely that landfill disposal of products containing tralomethrin represents a threat to human health.

No data on the anaerobic biodegradation of warfarin, pyriproxyfen, and MGK-326 were located. Pyriproxyfen has a large K_{oc} value and thus it is expected that even slight degradation over the plume migration period, incorporating both anaerobic biodegradation in the plume and aerobic biodegradation at the plume edges, will attenuate the concentration of this compound in groundwater such that its hazard quotient would be very low. At this time it is not possible to assess the potential for degradation of either warfarin or MGK-326 in landfill leachate or in groundwater. These compounds have a moderate ability to leach according to their K_{oc} values, and neither compound is readily biodegraded in aerobic screening tests (Brorson et al. 1994; Chemicals Inspection and Testing Institute 1991). MGK-326 may be susceptible to hydrolytic degradation under alkaline conditions (Mill et al. 1987).

Only the two inorganic compounds that are not biodegradable, zinc phosphide and boric acid, exceed the "safe" exposure definition (HQ > 1) based on the model results. Of the two, zinc phosphide is hydrolyzed at low to neutral pH values as might be expected in a landfill environment; when the decay rate in the model incorporates hydrolysis, hazard quotients estimated for this compound at pH values of 4 and 7 are equal to zero. The high hazard quotient reported for boric acid is for a source concentration of $1 \times 10^{+4}$ ppm, which is considerably higher than that expected from the disposal of boric acid-containing products. The low value is for a source concentration of 0.05 ppm. Boric acid is neither biodegradable or hydrolyzable; it is commonly found in the environment. Soil contains from 5 to 150 ppm boron naturally; boron concentrations in surface waters range from 0.001 to 0.1 mg/L (USEPA 1993b). While associated with reproductive effects from chronic exposure at high concentrations, boron is also a required nutrient for both plants and animals, including humans.

Propoxur and abamectin both have measurable estimated hazard quotients although neither compound exceeds the "safe" exposure definition. Hazard quotient values for propoxur approach the "safe" exposure definition under only the most conservative modeling conditions (low K_{oc} value, lowest available biodegradation rate constant value); these are worst case scenarios and even under these conditions this compound exhibits "safe" exposure levels. By taking the sum of the available hazard quotients from Table 9, and not incorporating the hazard quotients for the two inorganics and warfarin, a hazard index value of 0.32 is determined for this set of indoor household pesticides following disposal into a MSW landfill.

A lifetime cancer risk was also calculated for adults for the seven compounds in Table 7 with a published Q_1^* value (Table 10). This set includes all those compounds in Table 2 that have been identified as probable human carcinogens (EPA classification of B1 or B2). An action level of 10^{-6} has been published by EPA for carcinogens (USEPA 1995a). When loss of compound from biodegradation alone results in a lifetime cancer risk of less than 10^{-8}, the hydrolysis data are not presented. From the initial set of seven compounds, only propoxur exceeds the action level. At alkaline pH values, propoxur is expected to hydrolyze readily; however, at neutral to acidic pH this compound is fairly stable. Propoxur is unique among the compounds found in Table 2 because of its very high mobility (K_{oc} value of 29) and relatively slow biodegradation rate. Compounds such as this may require greater attention from regulatory officials. A total lifetime cancer risk estimate for adults, calculated as the sum of the lifetime cancer risk values, is estimated as $0-1.9 \times 10^{-6}$. This represents approximately zero to two excess cancers per 1,000,000 people from the ingestion of groundwater contaminated by disposal of this set of indoor household pesticides into a MSW landfill.

This risk characterization was completed using the most conservative estimations available, both during the modeling step and in the calculation of risk, to give a value that represents the upper limit of risk to a human exposed to leachate containing a mixture of indoor household pesticides. RfD and Q_1^* values were available for only a subset of compounds in Table 2, which included both leaching and nonleaching compounds as defined by the OPP and the CDFA. While RfD values were published for several compounds that exceeded or possibly exceeded the criteria for leaching potential (diazinon, tetrachlorvinphos, pro-

Table 10. Estimation of lifetime cancer risk from the ingestion of contaminated drinking water for selected compounds found in indoor household pesticides.

Compound	Lifetime cancer risk	
	Biodegradation	Hydrolysis
Dichlorvos	<1E-8	
Propoxur	<1E-8 to 1.76E-6	<1E-8 to 1.0E-4
Carbaryl	<1E-8	
Permethrin	<1E-8	
Fenoxycarb	<1E-8	
p-Dichlorobenzene	<1E-8 to 1.2E-7	0.12 to 29
MGK-326	<1E-8	

poxur, methomyl, boric acid, naphthalene, *p*-dichlorobenzene, warfarin), other compounds that also exceeded or possibly exceeded these criteria did not have available values (piperonyl butoxide, esbiothrin, limonene, linalool, petroleum distillates, DEET, citronella oil, MGK 326, cedarwood oil, pindone). Based on these values, however, the cancer and noncancer risks to public health resulting from disposal of indoor household pesticides are minimal.

E. Improved Standards for Solid Waste Landfills

In addition to physical and chemical characteristics of indoor household pesticides that make them generally unlikely to leach following disposal in a MSW landfill, and the low risk involved to human health if leaching does occur, reports of pesticides leaching from landfills are unlikely to be repeated. Virtually all operating landfills in the U.S. must be in compliance with the minimum standards for landfill siting and design established by the EPA under RCRA in 1980. Importantly, these standards were substantially strengthened in 1991 (USEPA 1991b), so that both operational and design standards for new MSW landfills and for lateral expansions of existing landfills operating after October 1993 will be nearly equivalent to the standards for hazardous waste landfills of that time (Anonymous 1990). An exemption from the design criteria (Subpart D) and groundwater monitoring (Subpart E) provisions was initially given to small landfill units where there was no evidence of groundwater contamination. These are defined as landfills accepting 20 t or less of MSW daily, based on a yearly average, that have no evidence of groundwater contamination and either are used by a community that is unable to reach a regional waste management facility for 3 consecutive months in the year or are found in a location that receives less than or equal to 25 in. of precipitation annually (USEPA 1991b). However, as of October 1997 these small landfill units, other than those found in either dry or remote locations, must comply with Subpart E (USEPA 1997f). Obligatory design standards for landfills accepting more than 20 t of MSW daily include either a performance-based design that requires that maximum contaminant levels are not exceeded in the upper aquifer, at a "point-of-reference" near the landfill, or an EPA-designated landfill design including a composite liner and leachate collection system. As a result of these newer design and operational standards for MSW landfills accepting Subtitle D waste, the potential for indoor household pesticides to leach from new landfills and contaminate groundwater has been further reduced. Operating landfills, still accepting waste after October 1993, are subject to the operational standards required for new landfills but are not required to change their design standards. However, the number of landfills in the U.S. is decreasing rapidly as older, smaller landfills are phased out and new, larger regional landfills designed with leachate and gas collection, as well as treatment systems, are built.

Table 11. List of technical and common names for pesticides discussed in this article.

CAS Registry number	Common name	IUPAC name
51-03-6	Piperonyl butoxide	5-[2-(2-Butoxyethoxy)ethoxymethyl]-6-propyl-1,3-benzodioxole
52-68-6	Trichlorfon	Dimethyl 2,2,2-trichloro-1-hydroxyethylphosphonate
62-73-7	Dichlorvos	2,2-Dichlorovinyl dimethyl phosphate
63-25-2	Carbaryl	1-Naphthyl methylcarbamate
67-97-0	Cholecalciferol	(3β-5Z,7E)-9,10-Secacholesta-5,7,10(19)-trien-3-ol
81-81-2	Warfarin	(RS)-4-Hydroxy-3-(3-oxo-1-phenylbutyl)coumarin
82-66-6	Diphacinone	2-(Diphenylacetyl)indan-1,3-dione
83-26-1	Pindone	2-Pivaloylindan-1,3-dione
113-48-4	MGK-264	N-(2-Ethylhexyl)-8,9,10-trinorborn-5-ene-2,3-dicarboximide
114-26-1	Propoxur	2-Isopropoxyphenyl methylcarbamate
121-21-1	Pyrethrin I	(Z)-(S)-2-Methyl-4-oxo-3-(penta-2,4-dienyl)cyclopent-2-enyl (1R)-trans-2,2-dimethyl-3-(2-methylprop-1-enyl)cyclopropane carboxylate
121-29-9	Pyrethrin II	(Z)-(S)-2-Methyl-4-oxo-3-(penta-2,4-dienyl)cyclopent-2-enyl (E)-(1R)-trans-3-(2-methoxycarbonylprop-1-enyl)-2,2-dimethylcyclopropane carboxylate
134-62-3	DEET	N,N-Diethyl-3-methyl-benzamide
136-45-8	MGK-326	Dipropyl 2,5-pyridinedicarboxylate
300-76-5	Naled	1,2-Dibromo-2,2-dichloroethyl dimethyl phosphate
333-41-5	Diazinon	O,O-Diethyl-O-2-isopropyl-6-methylpyrimidin-4-yl phosphorothioate
584-79-2	Allethrin	(RS)-3-Allyl-2-methyl-4-oxocyclopent-2-enyl (1R,3R;1R,3S)-2,2-dimethyl-3-(2-methylprop-1-enyl)cyclopropanecarboxylate
2921-88-2	Chlorpyrifos	O,O-Diethyl-O-3,5,6-trichloro-2-pyridyl phosphorothioate
3691-35-8	Chlorophacinone	2-[2-(4-Chlorophenyl)-2-phenylacetyl]indan-1,3-dione
4151-50-2	Sulfluramid	N-Ethylperfluoro-octane-1-sulfonamide
7696-12-0	Tetramethrin	Cyclohex-1-ene-1,2-dicarboximidomethyl(1RS,3RS;1RS,3SR)-2,2-dimethyl-3-(2-methylprop-1-enyl)cyclopropanecarboxylate
8003-34-7	Pyrethrin	(Z)-(S)-2-Methyl-4-oxo-3-(penta-2,4-dienyl)cyclopent-2-enyl (1R)-trans-2,2-dimethyl-3-(2-methylprop-1-enyl)cyclopropanecarboxylate
10453-86-8	Resmethrin	5-Benzyl-3-furylmethyl (1RS,3RS;1RS,3SR)-2,2-dimethyl-3-(2-methylprop-1-enyl)cyclopropanecarboxylate
16752-77-5	Methomyl	S-Methyl-N-(methylcarbamoyloxy)thioacetimidate
22248-79-9	Tetrachlorvinphos	(Z)-2-Chloro-1-(2,4,5-trichlorophenyl)vinyl dimethyl phosphate

Table 11. (Continued).

CAS Registry number	Common name	IUPAC name
22781-23-3	Bendiocarb	2,3-Isopropylidenedioxyphenyl methylcarbamate
26002-80-2	Sumithrin	3-(Phenoxybenzyl (*1RS,3RS;1RS,3SR*)-2,2-dimethyl-3-(2-methylprop-1-enyl)cyclopropanecarboxylate
27519-02-4	Z-9 Tricosene	(*Z*)-Tricos-9-ene
28434-00-6	Esbiothrin	(*S*)-3-Allyl-2-methyl-4-oxocyclopent-2-enyl (*1R,3R*)-2,2-dimethyl-3-(2-methylprop-1-enyl)cyclopropanecarboxylate
28772-56-7	Bromadiolone	3-[3-(4'-Bromobiphenyl-4-yl)-3-hydroxy-1-phenylpropyl]-4-hydroxycoumarin
40596-69-8	Methoprene	Isopropyl (*E,E*)-(*RS*)-11-methoxy-3,7,11-trimethyldodeca-2,4-dienoate
41096-46-2	Hydroprene	Ethyl (*E,E*)-(*RS*)-3,7,11-trimethyldodeca-2,4-dienoate
51630-58-1	Fenvalerate	(*RS*)-α-Cyano-3-phenoxybenzyl (*RS*)-2-(4-chlorophenyl)-3-methylbutyrate
52315-07-8	Cypermethrin	(*RS*)-α-Cyano-3-phenoxybenzyl (*1RS,3RS;1RS,3SR*)-3-(2,2-dichlorovinyl)-2,2-dimethylcyclopropanecarboxylate
52645-53-1	Permethrin	3-Phenoxybenzyl(*1RS,3RS;1RS,3SR*)-3-(2,2-dichlorovinyl)-2,2-dimethylcyclopropanecarboxylate
52918-63-5	Deltamethrin	(*S*)-α-Cyano-3-phenoxybenzyl(*1R,3R*)-3-(2,2-dibromovinyl)-2,2-dimethylcyclopropanecarboxylate
56073-10-0	Brodifacoum	3-[3-(4'-Bromobiphenyl-4-yl)-1,2,3,4-tetrahydro-1-naphthyl-4-hydroxycoumarin
66230-04-4	Esfenvalerate	(*S*)-α-Cyano-3-phenoxybenzyl(*S*)-2-(4-chlorophenyl)-3-methylbutyrate
66841-25-6	Tralomethrin	(*S*)-α-Cyano-3-phenoxybenzyl (*1R,3S*)-2,2-dimethyl-3-[(*RS*)-1,2,2,2-tetrabromoethyl]cyclopropanecarboxylate
67485-29-4	Hydramethylnon	5,5-Dimethylperhydropyrimidin-2-one 4-trifluoromethyl-α-(4-trifluoromethylstyryl)cinnamylidenehydrazone
68359-37-5	Cyfluthrin	(*RS*)-α-Cyano-4-fluoro-3-phenoxybenzyl (*1RS,3RS;1RS,3SR*)-3-(2,2-dichlorovinyl)-2,2-dimethylcyclopropanecarboxylate
71751-41-2	Abamectin	(*10E,14E,16E,22Z*)-(*1R,4S,5'S,6S,6'R,8R,12S,13S,20R,21R,24S*)-6"[(*s*)-sec-Butyl]-21,24-dihydroxy-5',11,13,22-tetramethyl-2-oxo-3,7,19-trioxatetracyclo[15.6.1.1^{4,8}.0^{20,24}]pentacosa-10,14,16,22-tetraene-6-spiro-2'-(5',6'-dihydro-2'*H*-pyran)-12-yl 2,6-dideoxy-4-*O*-(2,6-dideoxy-

Table 11. (Continued).

CAS Registry number	Common name	IUPAC name
		3-Oo-methyl-α-L-arabino-hexopyranosyl)-3-O-methyl-α-L-arabino-hexopyranoside mixture with ($10E,14E,16E,22Z$)-($1R,4S,5'S,6S,6'R,8R,12S,1$-$3S,20R,21R,24S$)-21,24-dihydroxy-6'-isopropyl-5',11,13,22-tetramethyl-2-oxo-3,7,19-trioxatetracyclo-11,13,22-tetramethyl-2-oxo-3,7,19-trioxatetracyclo[15.6.1.1^{4,8}.0^{20,24}]pentacosa-10,14,16,22-tetraene-6-spiro-2'-(5',6'-dihydro-2'H-pyran)-12-yl 2,6-dideoxy-4-O-(2,6-dideoxy-3-Oo-methyl-α-L-arabino-hexopyranosyl)-3-O-methyl-α-L-arabino-hexopyranoside (4:1)
72490-01-8	Fenoxycarb	Ethyl 2-(4-phenoxyphenoxy)ethylcarbamate
95737-01-8	Pyriproxyfen	4-Phenoxyphenyl (RS)-2-(2-pyridyloxy)propyl ether
120068-37-3	Fipronil	(±)-5-Amino-1-(2,6-dichloro-α,α,α—trifluoro-p-tolyl)-4-trifluoromethylsulfinylpyrazole-3-carbonitrile

Summary

Many indoor household pesticides are efficient and useful tools for a variety of functions necessary to maintain clean, sanitary, and pleasant homes and institutional facilities, and to provide significant public health benefits. They do so by incorporating active ingredients and formulation technology that have not been associated with significant environmental impact in use or when disposed in landfills. Chemical and environmental fate properties, toxicological characteristics, and use patterns of indoor household pesticides that distinguish them from other categories of pesticides which have been associated with environmental contamination should be recognized when HHW policy is debated and established by governmental agencies. Most indoor household pesticides as defined here should not be considered hazardous waste or HHW because those relatively few containers, often no longer full, that have been disposed with MSW over the years have not been associated with environmental contamination. The tiny amounts of those product residues that will reach MSW landfills have been shown, in general, not to have chemical or environmental fate characteristics that would make them susceptible to leaching. Those that do have the potential to leach based on these characteristics, in most cases, do not represent a threat to human health based on toxicological considerations. However, compounds such as propoxur, which are very mobile and relatively persistent in soil and in addition have been associated with significant potential health effects, may be targeted by the screening process as described here and could be selected for further investigation as candidates for special waste management status (such as HHW).

Our analysis and recommendations have not been extended to the many types of lawn and garden pesticides that are commonly used by homeowners and are frequently brought to HHW programs. However, their potential for groundwater contamination could also be judged using the same technical considerations as applied in this review to indoor household pesticides. In light of the very high costs of diverting wastes from the MSW stream and into HHW programs, it is recommended that, as a matter of public policy, all categories of household waste that might be considered as HHW be carefully and objectively evaluated for their potential to harm public health or the environment after disposal at MSW landfills.

References

Adcock JA, Bentley A, Challis I (1975) The decline of NC-6897 in a sandy loam soil. Report No. Metabolism 7512. Unpublished study received September 10, 1975 under 10065-3. CDL: 195089-D. Fisons Corporation, Agricultural Chemical Division, Bedford, ME.

Al-Bashir B, Cseh T, Leduc R, Samson R (1990) Effect of soil/contaminant interactions on the biodegradation of naphthalene in flooded soil under denitrifying conditions. Appl Microbiol Biotechnol 34:414–419.

Anonymous (1983) How hazardous are municipal wastes? Am City County March: 41–42.

Anonymous (1990) Stop hazwaste at the landfill. World Wastes February:18–19.

ATSDR (Agency for Toxic Substances and Disease Registry) (1996) Toxicological profile for diazinon (update). ATSDR, Atlanta, GA.

Barlaz MA, Ham RK (1993) Leachate and gas generation. In: Daniel DE (ed) Geotechnical Practice for Waste Generation, vol 6. Chapman & Hall, London, pp 113–136.

Bartkowiak D, Newhart K, Pepple M, Troiano J, Weaver D (1997) Sampling for pesticide residues in California well water: 1996 update of the well inventory database. EH 96-06. CA-EPA, Department of Pesticide Regulation, Sacramento, CA.

Bauer K, Garbe D, Surburg H (1988) Flavors and fragrances. In: Gerhartz W (ed) Ullmann's Encyclopedia of Industrial Chemistry, vol A11. VCH, New York, pp 141–239.

Bertrand H, Oliver D, Tormey M, Cearley D, Beck RW (1995) Household hazardous waste characterization study for Palm Beach County, Florida: a MITE program evaluation. EPA/600/R-95/140. National Risk Management Research Laboratory, Office of Research and Development, USEPA, Cincinnati, OH.

Binder RG, Flath RA (1989) Volatile components of pineapple guava. J Agric Food Chem 37:734–736.

Bjerg PL, Brun A, Nielsen PH, Christensen TH (1996) Application of a model accounting for kinetic sorption and degradation to in situ microcosm observations on the fate of aromatic hydrocarbons in an aerobic aquifer. Water Resour Res 32:1831–1841.

Block SS (ed) (1991) Disinfection, Sterilization, and Preservation, 4th Ed. Lea & Febiger, Malvern, PA.

Bobe A, Coste CM, Cooper JF (1997) Factors influencing the adsorption of fipronil on soils. J Agric Food Chem 45:4861–4865.

Boethling RS, Lynch DG (1992) Quaternary ammonium surfactants. In: Hutzinger O (ed) The Handbook of Environmental Chemistry, vol 3, part F. Springer-Verlag, Heidelberg, pp 146–177.

Bomberger DC, Lewis R, Valdes A (1988) Waste characterization study: assessment of recyclable and hazardous components. SRI Project #PYH-2530. SRI International for the California Waste Management Board, Sacramento, CA.

Bromilow R, Briggs G, Williams M, Smelt J, Tuinstra L, Traag W (1986) The role of ferrous ions in the rapid degradation of oxamyl, methomyl, and aldicarb in anaerobic soils. Pestic Sci 17:535–547.

Brorson T, Bjorklund I, Svenstam G, Lantz R (1994) Comparison of two strategies for assessing ecotoxicological aspects of complex wastewater from a chemical-pharmaceutical plant. Environ Toxicol Chem 13:543–52.

Budavari S, O'Neil MJ, Smith A, Heckelman PE (ed) (1996) The Merck Index. An Encyclopedia of Chemicals and Drugs, 12th Ed. Merck, Whitehouse Station, NJ.

Cahn A, Lynn JL Jr (1983) Surfactants and detersive systems. In: Grayson M (ed) Kirk-Othmer Encyclopedia of Chemical Technology, vol 22. Wiley, New York, pp 332–432.

Chapman RA, Cole CM (1982) Observation on the influence of water and soil pH on the persistence of insecticides. J Environ Sci Health B17:487.

Chemicals Inspection and Testing Institute (1991) Biodegradation and Bioaccumulation Data of Existing Chemicals Based on the CSCL Japan. Chemical Products Safety Division, Basic Industries Bureau, Ministry of International Trade and Industry, Tokyo, Japan.

Cubbage CP (1992) 1992 Update for state agricultural pesticide collections. In: Proceedings of the National U.S. EPA Conference on Household Hazardous Waste Management, Minneapolis, MN. NTIS PB93-170116. Prepared by Waste Watch Center, Andover, MA.

DHHS/NTP (1990) Toxicology and carcinogenesis studies of d-limonene (gavage studies). Tech rep ser 347. NIH Publ. No. 90-2802. National Institute of Health, Bethesda, MD, p 3.

Domsch KH (1984) Effects of pesticides and heavy metals on biological processes in soil. Plant Soil 76:367–378.

Ellington JJ, Stancil FE, Payne WD, Trusty CD (1988) Measurement of Hydrolysis Rate Constants for Evaluation of Hazardous Waste Land Disposal, vol 3. Data on 70 Chemicals. NTIS PB88-234 042/AS; EPA 600/S3-88/028. U.S. Environmental Research Laboratory, Athens, GA.

Fredrickson L, Latham C, Mitchell S, Thomas J (1992) Minnesota Pollution Control Agency solid waste composition study, 1990–1991, part I. Presented to the Legislative Commission on Waste Management by the Minnesota Pollution Control Agency, St. Paul, MN.

Gerike P, Gode P (1990) The biodegradability and inhibitory threshold concentrations of some disinfectants. Chemosphere 21:799–812.

Gianessi LP (1992) U.S. Pesticide Use Trends: 1966–1989. Prepared for USEPA's Office of Policy Analysis by Resources for the Future, Washington, DC.

Glenn J (1998) The state of garbage in America, part 1. BioCycle April: vol. 40, no. 4, 32–43.

Gomez E, Ledbetter CA, Hartsell PA (1993) Volatile compounds in apricot, plum and their interspecific hybrids. J Agric Food Chem 41:1669–1676.

Gonsior SJ, Bailey RE, Rhinehart WL, Spence MW (1984) Biodegradation of o-phenylphenol in river water and activated sludge. J Agric Food Chem 32:593–596.

Grenney WJ (1987) A mathematical model for the fate of hazardous substances in soil: model description and experimental results. Hazard Waste Hazard Mater 4:223–239.

Gump W (1979) Disinfectants and antiseptics. In: Grayson M (ed) Kirk-Othmer Encyclopedia of Chemical Technology, vol 7. Wiley, New York, pp 793–832.

Gustafson DI (1989) Groundwater ubiquity score: a simple method for assessing pesticide leachability. Environ Toxicol Chem 8:339–357.

Hansch C, Leo AJ (1981) Medchem Project. Issue 19. Pomona College, Claremont, CA.

Hansch C, Leo A, Hoekman D (1995) Exploring QSAR: Hydrophobic, Electronic, and Steric Constants. ACS Professional Reference Book, American Chemical Society: Washington, DC.

Hayes WJ, Lawes ER (eds) (1990) Handbook of Pesticide Toxicology, vol 3. Classes of Pesticides. Academic Press, New York.

Hill BD, Schaalje GB (1985) A two compartment model for the dissipation of deltramethrin on soil. J Agric Food Chem 33:1001–1006.

Hilton HW, Robison WH (1972) Fate of zinc phosphide and phosphine in the soil-water environment. J Agric Food Chem 20:1209–1213.

Howard PH (1989) Handbook of Environmental Fate and Exposure Data for Organic Chemicals: Large Production and Priority Pollutants, vol 1. Lewis, Chelsea, MI.

Howard PH, Meylan WM (1997) Handbook of Physical Properties of Organic Chemicals. CRC Lewis, Boca Raton, LA.

IARC (1987) Overall evaluations of carcinogenicity: an updating of IARC monographs, vol 1–42. IARC Monographs on the Evaluation of Carcinogen Risks to Humans. IARC, Lyon, France. WHO 7(suppl):56–74.

IARC (1991) Occupational exposures in insecticide application, and some pesticides. IARC Monographs on the Evaluation of Carcinogen Risks to Humans. IARC, Lyon, France. WHO 53:251–349.

Johnson B (1991) Setting revised specific numerical values, April 1991. Pursuant to the Pesticide Contamination Prevention Act. EH 91-6. California Department of Food and Agriculture, Environmental Monitoring and Pesticide Management Branch, Sacramento, CA.

Jordan ED, Kaufman DD (1986) Degradation of *cis-* and *trans-*permethrin in flooded soil. J Agric Food Chem 34:880–884.

Kanazawa J (1987) Biodegradability of pesticides in water by microbes in activated sludge. Environ Monit Assess 9:57–70.

Kawamota K, Urano K (1990) Parameter for predicting fate of organochlorine pesticides in the environment. III. Biodegradation rate constants. Chemosphere 21:1141–1152.

Kinman RN, Nutini DL (1988) Household hazardous waste in the sanitary landfill. Chem Times Trends July: vol. 11, 23–29, 39–40.

Kool JB, Huyakorn PS, Sudicky EA, Saleem AZ (1994) A composite modeling approach for subsurface transport of degrading contaminants from land-disposal sites. J Contam Hydrol 17:69–90.

Korotova LG, Denchenko AS (1978) Chlorophos degradation rate in chestnut soil and its transport by surface run-off water. Gidrokhim Mater 74:99–103.

Kuhn EP, Suflita JM (1989) Anaerobic biodegradation of nitrogen-substituted and sulfonated benzene aquifer contaminants. Hazard Waste Hazard Materials 6:121–133.

Lankas GR, Gordon LR (1989) Ivermectin and Abamectin. Springer-Verlag, New York.

Leistra M, Tuinstra LGMT, Vanderburg AMM, Crum SJH (1984) Contribution of leaching of diazinon, parathion, tetrachlorvinphos, and triazophos from glasshouse soils to their concentration in water courses. Chemosphere 13:403–413.

Lemmon CR, Pylypiw HM JR (1992) Degradation of diazinon, chlorpyrifos, isofenphos, and pendimethalin in grass and compost. Bull Environ Contam Toxicol 48:409–415.

Li J, Perdue EM (1995) Physicochemical properties of selected monoterpenes. Preprints of papers presented at the 209th ACS National Meeting, Anaheim, CA, April 2–7, 1995, 35:134–137.

Lyman WJ, Reehl WF, Rosenblatt DH (1990) Handbook of Chemical Property Estimation Methods. American Chemical Society, Washington, DC, pp 4–9.

Masunaga S, Susarla S, Yonezawa Y (1996) Dechlorination of chlorobenzenes in anaerobic estuarine sediment. Water Sci Technol 33:173–180.

Menzie CM (1972) Fate of pesticides in the environment. Annu Rev Entomol 17:199–222.

Meylan WM, Howard PH (1995) Atom/fragment contribution method for estimating octanol-water partition coefficients. J Pharm Sci 84:83–92.

Meylan WM, Howard PH, Boethling RS (1992) Molecular topology/fragment contribution method for predicting soil sorption coefficients. Environ Sci Technol 26:1560–1567.

Michel FC JR, Reddy CA, Forney LJ (1997) Fate of carbon-14 diazinon during the composting of yard trimmings. J Environ Qual 26:200–205.

Mill T, Haag W, Penwell P, Pettit T, Johnson H (1987) Environmental Fate and Exposure Studies Development of a PC-SAR for Hydrolysis: Esters, Alkyl Halides and Epoxides. EPA Contract No. 68-02-4254. SRI International, Menlo Park, CA.

Miller C, Pepple M, Troiano J, Weaver D, Kimaru W (1990) Sampling for pesticide residues in California well water: 1990 update well inventory data base. EH 90-11. California Department of Food & Agriculture, Environmental Hazards Assessment Program, Sacramento, CA.

Misra G, Pavlostathis SG, Perdue EM, Araujo R (1996) Aerobic biodegradation of selected monoterpenes. Appl Microbiol Biotechnol 45:831–838.

Moshonas MG, Shonas MG, Shaw PE (1994) Quantitative determination of 46 volatile constituents in fresh, unpasteurized orange juices using dynamic headspace gas chromatography. J Agric Food Chem 42:1525–1428.

Noble A (1993) Partition coefficients (n-octanol-water) for pesticides. J Chromatogr 642:3–14.

Ohkawa H, Nambu K, Inui H, Miyamoto J (1978) Metabolic fate of fenvalerate (sumicidin) in soil and by soil microorganisms. Nippon Noyaku Gakkaishi 3:129–41.

Park KS, Sims RC, Dupont RR, Doucette WJ, Matthews JE (1990) Fate of PAH compounds in two soil types: influence of volatilization, abiotic loss and biological activity. Environ Toxicol Chem 9:187–195.

Paulus W, Genth H (1983) Microbial phenolic compounds-a critical examination. In: Oxley TA, Barry S (eds) Biodeterioration. Papers presented at the 5th International Biodeterioration Symposium, Aberdeen, Scotland, pp 701–712.

Puget Sound Council of Governments (1985) Characterization and impacts of non-regulated hazardous waste in the solid waste of King County. CRS Report 1118-2, Seattle, WA.

Rathje WL (1998) Beyond the pail: the hazards of being a can-tosser. MSW Management, May/June.

Rathje WL, Wilson DC, Lambou VW, Herndon RC (1987) Characterization of household hazardous waste from Marin County, California, and New Orleans, Louisiana.

EPA 600/4-87-025, NTIS PB 88-102439, Report to the USEPA, Office of Research & Development. Las Vegas, NV.

Rogers IH, Birtwell IK, Kruzynski GM (1986) Organic extractables in municipal wastewater. Vancouver, British Columbia. Water Pollut Res J Can 21:187–204.

Ruckelshaus WD (1983) Science, risk, and public policy. Science 221:1026–1028.

Sack TM, Steele DH, Hammerstrom K, Remmers J (1992) A survey of household products for volatile organic products. Atmos Environ 26A:1063–1070.

Savage GM, Miller WL (1997) An investigation of the effects of household chemical products on methane production in landfill simulation laboratory reactors. Doc 1411. Prepared by CalRecovery, Inc., Hercules, CA and University of Florida, Gainesville, FL, for the Chemical Specialties Manufacturers Association, Washington, DC.

Shiu WY, Ma A, Mackay D, Seiber JN, Wauchope RD (1990) Solubilities of pesticide chemicals in water. Part II: Data compilation. Rev Environ Contam Toxicol 116: 15–187.

Smith RA (1985) Boric oxide, boric acid, and borates. In: Gerhartz W (ed) Ullmann's Encyclopedia of Industrial Chemistry. VCH, New York, pp A4:263–280.

Sumitomo Chemical Company, Inc. (1992) Degradation and environmental data package on sumithrin, neo-pynamin and pynamin forte. Sumitomo Chemical America, Inc., 345 Park Ave., New York, NY.

Suryanarayana MVS (1991) Structure-activity relationship studies with mosquito repellent amides. J Pharm Sci 80:1055–1057.

Tang J, Jun QZ, Shen GH, Ho CT, Chang SS (1983) Isolation and identification of volatile compounds from fried chicken. J Agric Food Chem 31:1287–1292.

Tellus Institute (1991) Disposal cost fee study final report. Rep 90-131. Prepared for the California Integrated Waste Management Board.

Tirey DA, Dellinger B, Rubey WA, Taylor PH (1993) The Thermal Degradation Characteristics of Environmentally Sensitive Pesticide Products. Risk Reduction Engineering Laboratory, Office of Research and Development, EPA/600/R-93/102, USEPA, Cincinnati, OH.

Tomlin CDS (ed) (1994) The Pesticide Manual, 10th Ed. British Crop Protection Council, Farnham, Surrey, UK.

Tomlin CD (ed) (1997) The Pesticide Manual, 11th Ed. British Crop Protection Council, Farnham, Surrey, UK.

U.S. Borax Company (1993) Material safety data sheet on boric acid. Valencia, CA.

U.S. Department of Agriculture Soil Conservation Service/ Agricultural Research Service Pesticide Properties Database (1997) http://wizard.arusda.gov/rsml/ppdb.html

U.S. Environmental Protection Agency (USEPA) (1979) Registration submission on brodifacoum. Environmental Fate Branch, Office of Pesticide Programs, Registration Division, USEPA, Washington, DC.

U.S. Environmental Protection Agency (USEPA) (1986) Survey of household hazardous wastes and related collection programs. EPA/530-SW-86-038. USEPA, Washington, DC.

U.S. Environmental Protection Agency (USEPA) (1988a) Criteria for municipal solid waste landfills. Summary of data on municipal solid waste landfill leachate characteristics. EPA/530-SW-88-038, NTIS PB 88-242441. USEPA, Washington, DC.

U.S. Environmental Protection Agency (USEPA) (1988b) Chemical fact sheet number 193 for resmethrin. Office of Pesticide Programs, Registration Division, USEPA, Washington, DC.

U.S. Environmental Protection Agency (USEPA) (1989) Risk assessment guidance for

Superfund, vol I. Human health evaluation manual, Part A. EPA/540/1-89/002. USEPA, Washington, DC.

U.S. Environmental Protection Agency (USEPA) (1990a) National survey of pesticides in drinking water wells. Phase I. EPA 570/9-90-015, NTIS PB91-125765. USEPA, Washington, DC.

U.S. Environmental Protection Agency (USEPA) (1990b) Pesticide fact sheet number 89. Avermectin B1. Office of Pesticide Programs, Registration Division, USEPA, Washington, DC.

U.S. Environmental Protection Agency (USEPA) (1990c) National oil and hazardous substances pollution contingency plan. CFR 40 Part 300. USEPA, Washington, DC.

U.S. Environmental Protection Agency (USEPA) (1991a) Criteria for classifying pesticides for restricted use due to groundwater concerns. Fed Reg 56:22076–22085.

U.S. Environmental Protection Agency (USEPA) (1991b) Criteria for solid waste disposal facility and criteria, Final Rule. 40 CFR Parts 257 and 258. USEPA, Washington, DC.

U.S. Environmental Protection Agency (USEPA) (1991c) R.E.D. facts sheet on methoprene. Office of Pesticide Programs, Registration Division, USEPA, Washington, DC.

U.S. Environmental Protection Agency (USEPA) (1991d) Integrated risk information system. DOE/EH-0194 (NTIS DE0101850). USEPA, Washington, DC (http://wwwepa.govngispgm3/iris/subst-fl.htm).

U.S. Environmental Protection Agency (USEPA) (1992a) Another look: national survey of pesticides in drinking water wells. Phase II. EPA 570/9-91-020, NTIS PB 92-120831. USEPA, Washington, DC.

U.S. Environmental Protection Agency (USEPA) (1992b) Pesticides in groundwater database. A compilation of monitoring studies: 1971–1991 national summary. EPA 734-12-92-001. USEPA, Washington, DC.

U.S. Environmental Protection Agency (USEPA) (1992c) Regulations for the acceptance of certain pesticides and recommended procedures for the disposal and storage of pesticides and pesticides containers. 40 CFR Part 165. USEPA, Washington, DC.

U.S. Environmental Protection Agency (USEPA) (1992d) Pesticide registration and classification procedures. 40 CFR Part 152. USEPA, Washington, DC.

U.S. Environmental Protection Agency (USEPA) (1993a) EPA's composite model for leachate migration with transformation products: EPACMTP, vol I. Background document. Office of Solid Waste, USEPA, Washington, DC.

U.S. Environmental Protection Agency (USEPA) (1993b) Reregistration eligibility decision (RED)—boric acid and its sodium salts. EPA-738-R-93-017. USEPA, Washington, DC.

U.S. Environmental Protection Agency (USEPA) (1994a) Integrated risk information system. RCRA/CERCLA Division, USEPA, Washington, DC, pp 3–21.

U.S. Environmental Protection Agency (USEPA) (1994b) Modeling approaches for simulating three-dimensional migration of land and disposal leachate with transformation products and consideration of water table mounding, vol II. Users guide for EPA's composite model for leachate migration with transformation products (EPACMTP). Office of Solid Waste, USEPA, Washington, DC.

U.S. Environmental Protection Agency (USEPA) (1995a) Hazardous waste management system: identification and listing of hazardous waste: hazardous waste identification rule (HWIR). Fed Reg 60:66344–66469.

U.S. Environmental Protection Agency (USEPA) (1995b) Reregistration eligibility decision (RED)—ortho-benzyl-para-chlorophenol. EPA-738-F-96-027. USEPA, Washington, DC.

U.S. Environmental Protection Agency (USEPA) (1996a) New pesticide fact sheet—fipronil. EPA-737-F-96-005. USEPA, Washington, DC.

U.S. Environmental Protection Agency (USEPA) (1996b) Merck Co., Inc. Notice of pesticide petition filing, December 10, 1996, Fed Reg 61:65043–65047.

U.S. Environmental Protection Agency (USEPA) (1996c) Office of pesticide programs list of chemicals evaluated for carcinogenic potential. Memorandum from Health Effects Division. USEPA, Washington, DC.

U.S. Environmental Protection Agency (USEPA) (1997a) Exposure factors handbook, vol 1: general factors. EPA/600/P-95/002FA. USEPA, Washington, DC.

U.S. Environmental Protection Agency (USEPA) (1997b) Fenoxycarb; pesticide tolerances for emergency exemptions. 40 CFR Part 180, April 25, 1997. Fed Reg 62: 20111–20117.

U.S. Environmental Protection Agency (USEPA) (1997c) (S)-Hydroprene biochemical pest control agent; pesticide tolerance. 40 CFR Parts 180 and 185, June 4, 1997. Fed Reg 62:30549–30554.

U.S. Environmental Protection Agency (USEPA) (1997d) Cyfluthrin; pesticide tolerances. 40 CFR Parts 180, 185, and 186, November 26, 1997. Fed Reg 62:63010–63019.

U.S. Environmental Protection Agency (USEPA) (1997e) Health effects assessment summary tables, FY 1997 update. EPA-540-R-97-036, PB97-921199. USEPA, Washington, DC.

U.S. Environmental Protection Agency (USEPA) (1997f) Revisions to criteria for municipal solid waste landfills; final rule and proposed rule. 40 CFR Part 258, July 29, 1997. Fed Reg 62:40707–40713.

U.S. Environmental Protection Agency (USEPA) (1998a) Deltamethrin; pesticide tolerance. 40 CFR Part 180, August 26, 1998. Fed Reg 63:45406–45414.

U.S. Environmental Protection Agency (USEPA) (1998b) Esfenvalerate; pesticide tolerance. 40 CFR Part 180, September 11, 1998. Fed Reg 63:48607–48615.

U.S. Environmental Protection Agency (USEPA) (1998c) Pyriproxyfen (2-[1-methyl-2-(4-phenoxypphenoxy)ethoxy]pyridine; pesticide tolerance. 40 CFR Part 180, July 6, 1998. Fed Reg 63:36366–36373.

U.S. Environmental Protection Agency (USEPA) (1998d) Reregistration eligibility decision (RED)—methomyl. EPA-738-R-98-021. USEPA, Washington, DC.

U.S. Environmental Protection Agency (USEPA) (1998e) Pesticides industry sales and usage: 1996 and 1997 Market Estimates. EPA-733-R-98-001. Office of Prevention, Pesticides, and Toxic Substances, USEPA, Washington, DC.

Vandervoort C, Zabik MJ, Branham B, Lickfeldt DW (1997) Fate of selected pesticides applied to turfgrass: effect of composting on residues. Bull Environ Contam Toxicol 58:38–45.

Waste Watch Center (1993) HHW Collection Programs in the U.S.–1993. Report available from the Waste Watch Center, 16 Haverhill Street, Andover, MA.

Waste Watch Center (1997) Regional reports. Available from the Waste Watch Center, 16 Haverhill Street, Andover, MA.

Wauchope RD, Butler TM, Hornsby AG, Augustijn-Beckers PWM, Burt JP (1991) The SCS/ARS/CES pesticide properties database for environmental decision-making. Rev Environ Contam Toxicol 123:1–36. (http://wizard.arusda.gov/rsml/ppdb.html)

Werner AF, Wilson JD (1982) What do you need to know about evaluating the environmental safety of household disinfectants. Soap Cosmet Chem Spec 58:30–32.

Werner AF, Taulli TA, Michel PR, Williams MA (1983) Estimation and verification of the environmental fate of O-benzyl-p-chlorophenol. Arch Environ Contam Toxicol 12:569–575.

Wilkerson MR, Kim KD (1986) The Pesticide Contamination Prevention Act: setting specific numerical values. EH 86-02. California Department of Food and Agriculture, Environmental Monitoring and Pest Management, Sacramento, CA.

Yalkowsky SH, Dannenfelser RM (1992) Aquasol database of aqueous solubility, Version 5. College of Pharmacy, University of Arizona, Tucson, AZ (PC version).

Zoecon Corporation (1989) Technical bulletin on methoprene.

Manuscript received May 28, 1999; accepted June 2.

Mercury Modeling to Predict Contamination and Bioaccumulation in Aquatic Ecosystems

M.C.B. Braga · G. Shaw · J.N. Lester

Contents

I. Introduction	69
II. Mercury in the Environment	70
A. Speciation of Mercury	70
B. Mercury Cycling	73
III. Mercury Bioconcentration, Bioaccumulation, and Biomagnification	79
IV. Modeling of Mercury in Aquatic Systems	80
A. Conceptual Background	80
B. The Role of Data Quality in Modeling Mercury	81
C. Existing Models of Mercury Cycling	82
D. Research and Modeling Requirements Concerning Mercury in Aquatic Systems	86
V. Conclusions	87
Summary	88
References	88

I. Introduction

The term *heavy metals* usually refers to those metals between atomic numbers 21 (scandium) and 84 (polonium) that occur either naturally or from anthropogenic sources in natural waters. Such metals constitute a potential environmental hazard because of their extensive use, toxicity, and widespread distribution (WHO 1991; Schnoor 1996).

The toxicity of heavy metals depends on both their concentration and the chemical form of the element in the environment. Forms that are almost totally insoluble pass through the human body without causing harm. However, for some heavy metals such as mercury, the most toxic form is that which has alkyl groups attached to the metal. Such compounds are soluble in animal tissues and

---

Communicated by George W. Ware

M.C.B. Braga·G. Shaw·J.N. Lester (✉)
T.H. Huxley School of Environment, Earth Sciences and Engineering/Imperial College of Science, Technology, and Medicine, RSM Building, Prince Consort Road, SW7 2BP, London, UK

M.C.B. Braga
Universidade Federal do Paraná, Brazil

pass through biological membranes, causing serious damage to the central nervous system (Wren et al. 1995).

Since the 1970s, when some of the most important ecological properties of mercury speciation in aquatic ecosystems were outlined (Wood et al. 1972; Wood 1973; Jensen and Jernelöv 1972; Fagerström and Jernelöv 1971, in Fagerström and Jernelöv 1972), scientific efforts have been focused on identifying regional and global sources of contamination (Lindqvist 1985; Meili 1988; Hudson et al. 1994; Hultberg et al. 1994; Johansson and Iverfeldt 1994; Watras and Bloom 1994; Kelly et al. 1995; Malm et al. 1995; Meuleman et al. 1995; Porvari 1995; Tsiros and Ambrose 1998). Over the years, and especially since the events at Minamata Bay, evidence has shown that environmental and health problems related to mercury contamination are still of great concern (Doi and Ui 1973; WHO 1990; Akagi 1995; Aula et al. 1995; Krabbenhoft 1996a). Many studies have demonstrated mercury pathways in natural waters, soil, and in the atmosphere (Gavis and Ferguson 1972; Jernelöv and Åsell 1973; Brosset 1981; Lindqvist 1985; Jernelöv 1994b; Morrison and Theirén 1994; Munthe 1994; Malm et al. 1995; Porcella 1994; Allan and Heyes 1998; Scherbatskoy 1998), but much still must be learned about specific migration routes and mechanisms in environmental media.

Very little information is available on the key environmental factors that regulate the formation, destruction, and trophic transfer of methylmercury; the relative magnitude of natural versus anthropogenic sources of mercury; and detailed inventories of mercury emissions and regulation of emissions based on existing cleanup technologies (WHO 1990; Watras et al. 1994; Krabbenhoft and Rickert 1995; Krabbenhoft 1996a). In response to the specific environmental behavior of mercury and its inorganic and organic compounds, numerous studies have been undertaken to define the factors that can influence its bioconcentration and bioaccumulation (Meili 1988; Wang et al. 1989; USEPA 1990; Bubb et al. 1991; Zillioux et al. 1993; Jernelöv 1994a,b; Watras et al. 1994; Aula 1995; Guimarães 1995; Kelly et al. 1995; Meuleman et al. 1995; Porvari 1995; Allan and Heyes 1998; Balogh et al. 1998; Scherbatskoy et al. 1998). This article reviews some of these studies, specifically, the use of models to predict mercury concentration and bioaccumulation in aquatic systems, together with available information on mercury speciation and cycling in surface waters.

II. Mercury in the Environment
A. Speciation of Mercury

Speciation of metals to a large extent determines their toxicity. The term *speciation* refers to the determination of individual physicochemical forms of an element that together make up its total concentration in a sample. In particular, *physical speciation* involves differentiation of the physical size or physical properties: dissolved, colloidal, and particulate forms. *Chemical speciation* entails differentiation between the various chemical forms: elemental, organic, and inorganic species in the case of mercury. Chemical speciation also affects the relative degree of adsorption or binding to particles in natural waters, which affects its fate: sedimentation, precipitation, volatilization, and toxicity (Morel and Hering 1993).

In river and lake waters, mercury can be present in one or more of three different oxidation states: Hg^0 (metallic), Hg_2^{2+} (mercurous), and Hg^{2+} (mercuric) (Hem 1970). Mercury compounds differ greatly in their solubilities. Solubility in water increases in the following order: elemental mercury (2 μg L^{-1} at 30 °C) < mercurous chloride (2 mg L^{-1} at 25 °C) < methylmercury chloride < mercuric chloride (69 g L^{-1} at 20 °C). The solubility of methylmercury chloride is higher than that of mercurous chloride by about three orders of magnitude, this being related to the high solubility of methylmercury ion in water (WHO 1990).

The mercurous and mercuric states form numerous inorganic and organic chemical compounds that differ considerably in their toxic effects, although all are hazardous. Nonetheless, organomercury compounds are much more toxic. They tend to be lipid soluble and, because higher organisms do not decompose and excrete them effectively, tend to be concentrated in organic tissues in the food chain (Greeson 1970; WHO 1991).

According to Lindqvist and Rodhe (1985) the speciation of mercury compounds can be pictured as follows (Fig. 1):

V (volatile): Hg^0; $(CH_3)_2Hg$
R (water soluble or particle-borne): Hg^{2+}; HgX_2; HgX_3^-, and HgX_4^{2-} (X = OH^-, Cl^-, or Br^-); HgO on aerosol particles; Hg^{2+} complexes with organic acids
Water soluble: $HgCl_2$; $Hg(OH)_2$; CH_3HgCl
NR (nonreactive): CH_3Hg^+; CH_3HgCl; CH_3HgOH, and other organomercuric compounds; $Hg(CN)_2$; HgS, and Hg^{+2} bound to sulfur in fragments of humic matter.

The nature of the species that will occur in a specific aquatic system or predominate in solution is now well established (Hem 1970; Fargerströn and Jernelöv 1972; Gavis and Ferguson 1972; Schroeder 1988; Jernelöv 1994a), and

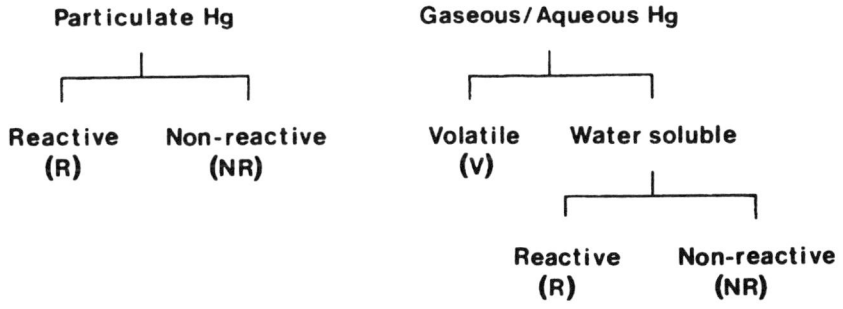

Fig. 1. Classification scheme for mercury species in the environment (Lindqvist and Rhode 1985).

depends on the redox potential (Eh), pH, and the nature of the anions and other groups present with which mercury can form stable complexes. In aqueous systems, where the pH is likely to fall to between 5 and 9 and measured Eh values are seldom higher than 0.5 V, Hg^0 and HgS are the species most likely to enter into equilibrium with mercury species in solution. At low redox potentials, observed in reducing media, mercury is immobilized by the sulfide ion. Although reducing conditions increase its solubility slightly, at pH values above 9 the solubility increases markedly, with the formation of HgS_2^{2-} ions (Gavis and Fergusson 1972; Björnberg 1988). It has been hypothesized by Fitzgerald et al. (1994) that the nonreactive mercury fraction present in deposition can be solubilised under anoxic conditions or in the presence of sulfite to yield species such as $Hg(SH)_2^0$, which can be bacterially methylated.

According to Fargerströn and Jernelöv (1972) the presence of mercuric sulfide, which possesses an extremely low solubility product (10^{-53}), can provide Hg^{2+} for methylation. For the bivalent mercury to be available for methylation, however, oxidation of the sulfur group from sulfide to sulfate must take place, as follows:

$$HgS + 2\ O_2 \rightarrow Hg^{2+} + SO_4^{2-}$$

Oxidation may occur at a very low rate as a physicochemical process in aerobic water, dependent on the redox potential. It can also be biologically mediated, leading to a somewhat faster release of bivalent mercury. However, the oxidation of sulfide is the rate-determining step in the two-step process.

$$HgS \rightleftharpoons Hg^{2+} \rightleftharpoons CH_3Hg^+ \text{ (Jernelöv 1994a)}$$

The methylmercury ion exists in aqueous solutions as an aquo-complex $CH_3\text{-}Hg\text{-}OH_2^+$ with a covalent bond between Hg and O. The cation behaves as a soft acid and has a strong preference for the addition of only one ligand. CH_3Hg^+ undergoes rapid coordination reactions with S, P, O, N, halogens, and C; the rate of formation of Cl^-, Br^-, and OH^- complexes is extremely fast. The CH_3Hg^+ itself, however, is kinetically inert and not amenable to decomposition. Therefore methylmercury compounds, once formed, are not readily demethylated. The neutral CH_3Hg^+ species are soluble at the low concentrations at which they are formed, and thus they are able to escape into solution. They are also lipophilic and can therefore readily pass through biological membranes. This fact, together with the tendency of the species to form stable complexes and the robustness of the CH_3Hg^+ unit, characterizes some of the far-reaching toxicological properties of methylmercury (Stumm and Morgan 1995).

Another type of mercury compound that is of importance in aquatic chemistry is dimethylmercury [$(CH_3)_2Hg$], in which mercury is attached to two carbon atoms by means of covalent bonds. This compound is considered insoluble, volatile, and remains undissociated in solution. Due to its volatility, dimethylmercury escapes from aqueous systems and is released into the atmosphere. Once in the atmosphere, it is converted to elemental mercury through photochemical reactions and eventually returns to the surface by deposition (Lindqvist and Rodhe 1985).

Hem (1970) used the free energy of formation of 33.5 kcal/mole for liquid dimethylmercury in the calculations to prepare the Eh–pH stability diagram for soluble mercury species in equilibrium in aqueous media. He concluded that there was no region in the diagram where $(CH_3)_2Hg$ could be the most stable phase. A comprehensive review of all the available literature has not revealed a free energy value for CH_3Hg^+ or any firm basis for calculating or estimating such a value.

B. Mercury Cycling

A crucial determinant of the potential effects of mercury in aquatic ecosystems is the cycling of its various species. *Cycling* consists of inputs to and outputs from a system, as well as transport, and transformation of mercury species within the system (Zillioux et al. 1993). These processes are affected by site-specific factors such as water chemistry, sediment transport and hydrodynamics (Johansson and Iverfeldt 1994; Gbondo-Tugbawa and Driscoll 1998). Mercury is introduced into the environment by a variety of natural processes and anthropogenic activities. Once mobilized, it can participate in the general circulation of matter throughout the hydrological cycle.

It is extremely difficult to estimate the abundance of mercury in the Earth's crust. Taylor (1964) pointed out that the mean crustal concentration of mercury was 0.08 ppm. Mercury is known to be transported by water from locations where it occurs in relatively high concentrations as deposits of cinnabar (HgS) and metallic mercury, and it is also carried through the atmosphere (Fleisher 1970). Its unusually high volatility, represented by its vapor pressure of 0.16 Pa or 1.2×10^{-3} mmHg (Jenne 1970; WHO 1991), accounts for its presence in the atmosphere. Fleisher (1970) pointed out that natural pollution caused by the volatility of mercury from its ore deposits or base metals could give values up to 62 ng m^{-3}. According to Lindqvist and Rodhe (1985), a dominant fraction of the total mercury in the atmosphere ($\approx 80\%$) consists of volatile gaseous mercury, probably elemental mercury. This mercury has an atmospheric residence time of between a few months and 2 years and is uniformly distributed throughout the atmosphere (1–2 ng m^{-3}). The few measurements that have been taken of the water-soluble fraction of mercury in the background air indicate concentrations of 0.005 to 0.2 ng m^{-3} (Brosset 1984, in Lindqvist 1985). It has been suggested by Fitzgerald et al. (1994) that reactive mercury found in precipitation and atmospheric particulate matter is derived from the atmospheric oxidation products of elemental mercury in the atmosphere. Thus, when it enters aqueous systems it is readily available to participate in competitive reactions associated with methylation, reduction to elemental mercury, uptake by biota, and sequestering with dissolved organic carbon. Once in surface waters, mercury enters a complex cycle (Fig. 2) in which one form can be converted to another. In gen-

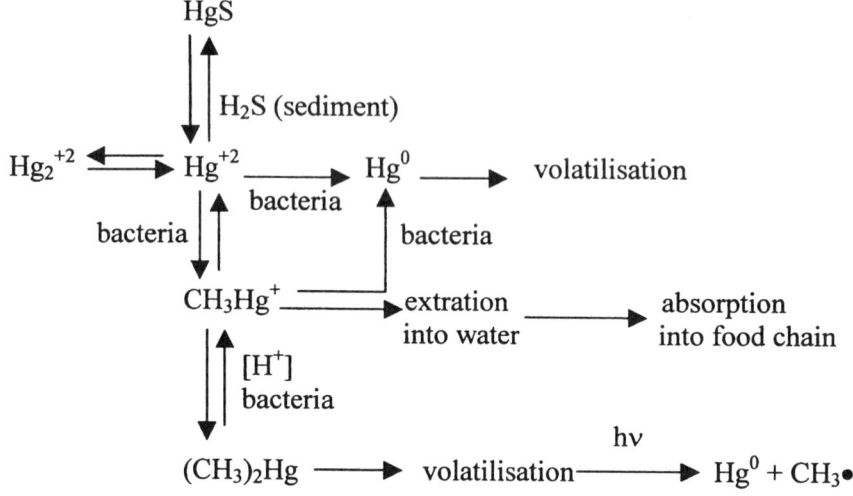

Fig. 2. General mercury cycle (adapted from Wood 1973, Bailey 1978, Brosset 1981, and Jernelöv 1994a).

eral, concentrations in surface waters tend to be very low, 1–3 ng L^{-1} (Lindqvist and Rodhe 1985).

Because of its volatility mercury can be released from the oceans, which it enters as a natural result of soil erosion, by volatilization. It subsequently, reaches the earth surface in both dry and wet deposition and is captured by the soil or enters the runoff and becomes part of the mercury content of surface natural water systems.

Before the 1980s, concerns about environmental contamination caused by the presence of mercury compounds in aquatic systems were focused on point-source discharges such as mining, smelting, and industrial or wastewater treatment plants. Technological solutions to many of these problems of mercury emission have been found. However, the problem of residual mercury within the environment still persists (Bubb et al. 1991). Recently, research has been conducted into mercury contamination based on non-point sources from atmospheric transport, thought to result from the combustion of fossil fuels and emissions from gold mining. There have been reports of contamination of lakes far from human influence, biota, and reservoirs for hydroelectric development, which were found to have high levels of mercury (Cossa et al. 1994; Lindqvist 1994; Verta et al. 1994; Watras et al. 1994; Mattice et al. 1997).

Research is being carried out in the Brazilian Amazon (Akagi et al. 1995; Guimarães et al. 1995; Malm et al. 1995; Porvari 1995) into point-source and non-point-source mercury contamination resulting from gold mining. Mercury enters the environment in the Amazon region either through the sublimation of mercury from amalgamation during burning or reburning or through direct release into aquatic systems. In the first scenario, mercury volatilizes into the air as metallic mercury; this elevates its background concentration in the atmosphere because of the long residence time of this element in air. If high humidity is present in the atmosphere, as in the Amazon, the oxidizing reaction (Hg^0 to Hg^{2+}) may be very quick, which reduces the residence time to a few days (Lindqvist 1985). Thus, the mercury emitted as mercury vapor will return rapidly to the adjacent forest as bivalent mercury and will eventually reach the aquatic system via runoff. The second scenario is related to metallic mercury that has been discharged into the water body during gold mining, deposited in the sediment, and partly oxidized to bivalent mercury.

This research demonstrated that 92% of the samples of predatory fish in the Tucurui reservoir, which is located in northeastern Brazilian Amazon, was contaminated by mercury. The level of contamination was higher than that established as the safe limit by the Brazilian Legislation for human consumption (0.5 µg Hg g^{-1} wet weight); for example, the mean mercury concentrations for three different species of predatory fish (Piranha, Tucunaré, and Pescada) were 2.6 mg Hg kg^{-1}, 1.1mg Hg kg^{-1}, and 1.2 mg Hg kg^{-1} wet weight, respectively (Porvari 1995). It was found by Aula et al. (1995) that direct emissions of mercury from the gold mining areas to the upstream tributaries of the Tucurui reservoir, together with the mercury load from soil erosion, are the probable cause of the high burden of mercury in the reservoir. Long-distance transport of mercury was verified by Malm et al. (1995) after assessing mercury contamination in different gold mining areas in the Amazon region. High mercury concentrations in fish were detected both 180 and 500 km away from point sources. Salati (1985, cited in Porvari 1995) supports the conclusion that mercury can be transported by the prevailing winds from a gold field located east of Tucurui.

Another example of airborne mercury load has been provided by Cossa et al. (1994), who studied Pavin Lake in France, a mountain lake remote from anthropogenic mercury sources. They concluded that the main route by which mercury entered the lake was atmospheric deposition, mostly as dissolved species. The high level was identified by means of the vertical profile of total dissolved mercury concentration of 9.1 pM in surface waters and 6 pM in the interface between the oxic and anoxic regions (about 60 m).

Meuleman et al. (1995) pointed out that atmospheric values measured during winter, in Lake Baikal in Russia, ranged from 2.9 ± 1.6 ng m^{-3} and 0.052 ± 0.025 ng m^{-3} for total gaseous mercury and total particulate mercury, respectively. These concentrations were slightly higher than in summer, when total gaseous mercury was found to be 1.45 ± 0.44 ng m^{-3} and total particulate mercury was 0.011 ± 0.005 ng m^{-3}. This result led the researchers to the conclusion that these higher concentrations were probably caused by three main factors: coal burning, vehicle traffic, and prevailing winds. Seasonal variation in the distribution of mercury species was also identified by Hurley et al. (1994), Verta et al. (1994), Watras et al. (1994), Jacobs et al. (1995), Bigham and Vandal (1996), and Benoit and Babiarz (1995, in Balogh et al. 1998).

Mercury may also enter the aquatic system in the insoluble form adsorbed onto suspended solids in runoff or in streams. Thus, it can be carried to the sediments by particle settling (Fitzgerald et al. 1994; Johansson and Iverfeldt 1994; Sherbatskoy et al. 1998; Cossa et al. 1994). If reducing conditions prevail, the presence of sulfide ion effectively immobilizes inorganic mercury, regardless of the chemical state of the mercury when it enters the sediment. However, the redox potential of the upper layer of the sediment may not be low enough to keep sulfur in the sulfide state. Hence, mercury can be released into solution by means of partial dissolution of manganese and iron oxides present in the sediment (Jenne 1970; Jacobs et al. 1995). Results by Watras and Bloom (1994) and Hurley et al. (1994) indicated that distributions of Fe and Mn were related to redox-driven diffusion of mercury from sediments into the overlying water column.

Wang et al. (1989) studied Katepwa Lake in Canada and concluded that dissolution of iron and manganese oxides was a significant mechanism in the release of mercury from the sediment to the pore water when dissolved oxygen was sufficiently depleted. It was found that substantial amounts of iron and manganese were dissolved in the solution of samples upon the depletion of oxygen. After periods of 1 hr and 96 hr, the resulting concentrations for Fe, Mn, and Hg were as shown in Table 1.

The mercury thus released becomes available to bacteria that methylate it to CH_3Hg^+ (see Fig. 2). Methylated mercury escapes into the water by diffusion and is dispersed by turbulence and convective circulation. Cossa et al. (1994) suggested that a high concentration of methylmercury at the interface between the oxic and the anoxic layers could be a result of either *in situ* production or accumulation of sinking particulate matter. It was shown that the particulate methylmercury concentration averaged 3.5% ± 2% of the particulate mercury concentration in the upper 40 m, whereas in the waters deeper than 60 m it was 2.8% ± 1.3%. They concluded that methylmercury could be formed at the interface.

Studies by Watras and Bloom (1994) of three seepage lakes in Wisconsin, U.S., concluded that total mercury and methylmercury concentrations in surface waters generally ranged from about 0.5 to 5 ng L^{-1} and 0.05 to 0.5 ng L^{-1}, respectively. However, much higher concentrations of both mercury and methylmercury were observed at depth, especially in plankton layers, reaching maxi-

Table 1. The release of Fe, Mn, and Hg from Katepwa Lake sediment as influenced by the dissolved oxygen levels at 25 °C.

Reaction period (hr)	pH	Eh (mV)	µmol released/kg sediment		
			Fe	Mn	Hg
1	7.75	11	10 ± 5	100 ± 20	27 ± 3
96	7.65	−35	150 ± 30	570 ± 90	68 ± 7

Source: Adapted from Wang et al. (1989).

mum concentrations of 45 and 11.6 ng L^{-1} for mercury and methylmercury, respectively. Mass balances for methylmercury indicated that in-lake production is an important source of methylmercury species and that the maximum concentration of waterborne methylmercury occurred near the sulfate–sulfide transition zone. The authors hypothesized that (a) methylmercury is formed *de novo* within plankton layers, perhaps via the same mechanisms that govern mercury methylation in sediments; and (b) a common mechanism governs *in situ* methylmercury production in both sediments and the water column.

Results from mercury speciation carried out in Lake Baikal by Meuleman et al. (1995) support the previous conclusion. It was shown by these authors that the methylmercury gradient increased with depth, and coincided with low oxygen and inorganic labile mercury levels. They concluded that there was *in situ* production of methylmercury under lower oxygen concentrations, below 150 m in the southern basin of the lake, where labile methylmercury made up 7%–15% of the total mercury.

Because of its high solubility in lipids relative to its solubility in water, mercury can enter the food chain or can be released back to the atmosphere by volatilization. Figure 2 shows that mercury volatilizes either as a result of methylmercury demethylation or methylation by bacterial action to form Hg^0 or $(CH_3)_2Hg$, respectively. However, owing to the bioaccumulation of methylmercury, methylation is more prevalent in the aquatic environment than demethylation.

According to the results of a survey carried out in aquatic systems in the Brazilian Amazon near gold mining areas (Guimarães et al. 1995), methylation was not detectable in water, but rates in surface sediments ranged from 10^{-5} to 10^{-1}% g^{-1} hr^{-1}. The highest values were found in a reservoir area (Samuel reservoir, southwestern Amazon region), with average values of 0.66% g^{-1} hr^{-1} (500 m downstream from the reservoir) and 0.69% g^{-1} hr^{-1} (flooded forest) at depths of 1 and 2 m, respectively. However, the rate found for the anoxic sediment collected 100 m from the dam, at a depth of 20 m, was 0.003% g^{-1} hr^{-1}. The authors concluded that this might result from the formation of ^{203}HgS, which is less susceptible to methylation than Hg^{2+}.

Net methylmercury production is dependent on the balance between methylation and demethylation processes influenced in a complex and variable manner by an array of biological and physicochemical parameters such as pH, temperature, dissolved organic carbon, oxygen concentration, mercury and methylmercury concentration, and availability (Fitzgerald et al. 1994), as well as by sulfate, manganese, and iron concentration, as already described.

Fargeström and Jernelöv (1972) summarized conclusions from previous work on the methylation of mercury and reported that the rate of methylation was well correlated with the general microbiological activity of organisms from different groups such as anaerobic and aerobic bacteria and fungi. Thus, the rate of methylation increases with a rise in temperature and with increased nutrient concentration of the substrate (Zillioux et al. 1993; Hurley et al. 1994; Kelly et al. 1995; Bigham and Vandal 1996; Balogh et al. 1998). Microorganisms metabolize mercury when it is present in their food in the form of particulate or dissolved organic matter (dissolved organic carbon, DOC). DOC is an effective complexing ligand for many trace metals, including mercury (Gilmour and Henry 1991; Morel and Hering 1993). In general, there are increased levels of mercury within aquatic systems if they contain elevated amounts of organic matter such as humic substances, represented generically as DOC (Lindqvist and Rodhe 1985; Verta et al. 1994; Watras et al. 1994; Allan and Heyes 1998; Balogh et al. 1998; Lodenius 1987 and Lindqvist et al. 1991, in Downs et al. 1998). The concentration of dissolved organic matter in waters is typically in the range of 1 to 100 mg L^{-1}. Typical average values are 1 mg L^{-1} in groundwater, 2–10 mg L^{-1} in lakes and rivers, and 10–50 mg L^{-1} in bogs and marshes (Morel and Hering 1993). In Sweden, a study of watersheds from three geographically separate areas demonstrated that the main factors controlling quantities of mercury in the runoff are the content of humic matter in the aquatic environment and the mercury concentrations in the humic substances (Johansson and Iverfeldt 1994).

Recent studies carried out in the U.S. and Sweden have demonstrated strong correlation between DOC and the total and methylmercury concentration in aquatic ecosystems (Meili 1988; Krabbenhoft and Babiarz 1992; Zillioux 1993; Hultberg 1994; Watras and Bloom 1994). Krabbenhoft et al. (1995) reported that the increase in mercury species was concomitant with increases in DOC. During the survey period the wetland pond water presented a variation in total mercury concentration from 0.9 to 6.3 ng L^{-1}, whereas methylmercury varied from 0.21 to 0.33 ng L^{-1}, in relation to the variation of DOC from 3.7 to 11.0 mg L^{-1}.

Jernelöv (1994a) pointed out that the rate of methylation activity is higher under anaerobic conditions than under aerobic, which agrees with results by Watras and Bloom (1994). If the conditions in a water body are examined, however, hydrogen sulfide is formed when oxygen is depleted, and this results in the formation of mercuric sulfide. Under anaerobic conditions no methylation occurs from mercuric sulfide. Therefore, in fresh water the methylation rate is likely to be higher under suboxic than under anaerobic conditions, as highlighted by Bubb et al. (1991), and Krabbenhoft (1996a), whose results agree with those of Bigham and Vandal (1996).

Another important factor that affects methylation–demethylation is the pH. Although pH ranging from 5 to 9 presents little influence on the overall rate of methylation, the relationship between monomethylmercury and dimethylmercury is strongly affected by pH. At higher pH values almost all methylated mercury formed is released as dimethylmercury, whereas at lower values monomethylmercury is the product (Jensen and Jernelöv 1972; Meili 1994; Krabbenhoft 1996a). Watras et al. (1994) pointed out that average waterborne mercury and methylmercury concentrations correlated negatively with lakewater pH.

The oxidation–reduction and methylation–demethylation reactions are assumed to be widespread in the environment, and each ecosystem attains its own steady state with respect to the individual species of mercury.

III. Mercury Bioconcentration, Bioaccumulation, and Biomagnification

The accumulation of toxic substances in organisms and in the food chain is an important guide in the assessment of the harmfulness of a substance in the environment. *Bioconcentration* can be defined as the direct uptake of aqueous toxicants through the gill membrane and epithelial tissue of organisms from the dissolved phase. It is part of the greater process of bioaccumulation and biomagnification, which includes effects on the food chain. *Bioaccumulation* refers to the total biouptake of a toxicant by an organism from a number of different sources, as well as via transport of dissolved matter through the gill and epithelium, whereas *biomagnification* describes a progressive accumulation of a toxicant in the food chain (Hartung 1994a; Schnoor 1996).

Despite the uncertainties surrounding speciation, it is known that the change from inorganic to methylated forms is the first crucial step in the aquatic bioaccumulation processes. In spite of the extremely low levels of mercury found in natural waters, bioaccumulation occurs because mercury, when methylated, enters the food chain by rapid diffusion and tight binding to proteins in aquatic biota (Wood et al. 1972). Methylmercury is rapidly accumulated by most aquatic biota and attains its highest concentration in fish at the top of the aquatic food chain, with a retention time of several years (WHO 1990).

Meili (1988) observed that mercury concentration in fish in Lake Blacksästjärn in Sweden exceeded the concentration in zooplankton by roughly one order of magnitude. The mean concentration in muscle tissue ranged from 2.1 to 2.7 mg kg^{-1} (μg g^{-1}) on a dry weight basis. The maximum value was observed in spring after the ice breakup and the minimum value in late summer. These results led the author to conclude that the decrease in mercury concentration in summer was probably a result of increasing body weight rather than a decrease in mercury content, i.e., biological dilution caused by enhanced growth. It has been suggested by Hurley et al. (1994) that increase in the mercury content in fish is related to increased levels of methylmercury in particulate matter, such as plankton, in lakes, which emphasizes the potentially important role of trophic level transfer in methylmercury bioaccumulation.

Porcella (1994) demonstrated that fish were the dominant methylmercury pool in the water column in the Little Rock Lake Treatment Basin, a seepage lake in Wisconsin, in the U.S., and that biomagnification of methylmercury progressed in the order phytoplankton (10^5) < zooplankton ($10^{5.5}$) < fish ($10^{6.5}$). These results were presented as a bioaccumulation factor and represented 15%, 30% and 95% as methylmercury in each trophic level.

Malm et al. (1995), assessing the Amazon River basin in Brazil, reported that the methylmercury concentrations in fish also varied depending on the trophic level. They studied more than 25 species but sampling was concentrated on the carnivorous species in recognition of their position at the top of the aquatic food chain. Results showed that the average methylmercury concentrations were 0.13 μg Hg g^{-1} for omnivorous, 0.18 μg Hg g^{-1} for microfagous, and 0.70 μg Hg g^{-1} for carnivorous fish, considered on a wet weight basis. These data confirmed that mercury ultimately accumulates in fish bodies by biomagnification in aquatic systems.

The older an organism is, and the higher its position in the food chain, the greater the fraction of total mercury existing as methylmercury. Nearly all (95%–100%) of the mercury present in fish is methylmercury, obtained mostly from the diet. As a consequence, those organisms that remain in the same body of water exhibit higher mercury concentrations and will return the mercury in their tissues to the mercury pool in the sediment when they die. In contrast, species that migrate excrete some of the mercury from their bodies as a result of discontinuity of inputs (Gavis and Ferguson 1972; Porcella 1994; Krabbenhoft 1996a; Krabbenhoft 1996b).

IV. Modeling of Mercury in Aquatic Systems
A. Conceptual Background

Water quality modeling has evolved significantly since the first mathematical model was developed by Streeter and Phelps in 1925 to assess organic contamination in the Ohio River in the U.S. (Chapra 1997). The 1960s brought changes in the way in which models were applied, especially as the digital computer allowed a more detailed approach to numerical problems related to water quality. During the 1970s, modeling studies were devoted to eutrophication, a phenomenon related to excessive growth of algae in surface waters caused by the high concentration of nutrients, nitrogen and phosphorus. The major development during the 1980s was the recognition of the important role of solids, suspended and settled, on the transport and fate of toxic substances, because of their role in bioaccumulation and biomagnification (Bubb et al. 1988; Gilmour and Henry 1991; Johansson and Iverfeldet 1994). In addition to the technical advances in modeling of water quality there has been a recognition that the protection of the environment is a crucial factor in the maintenance of the quality of life through the understanding and prediction of environmental processes and events.

The first steps in the modeling process are model identification and selection, the goals of which are to identify the simplest conceptual model that includes all the important phenomena which affect water quality and thus to select the most useful analytical formulae or computing codes. During model identification, available information is gathered and organized to construct a coherent picture of the water quality problem. The goal of model selection is to obtain a simulation model that effectively implements the conceptual model identified for each specific aquatic environment (USEPA 1990).

The level of sophistication of water quality models varies from what is known as a simple "box" model, which can be used to represent aquatic systems from one- to three-dimensional physically continuous models. The required sophistication of any model depends not only on the level at which the system is understood but also on how the model will be used.

During the last few years, mathematical models have varied widely in their ability to predict successfully the accumulation and transport of pollutants in environmental media. As a result, water quality models have recently become more sophisticated in their representation of chemical processes in aquatic systems and have incorporated more details on water chemistry and the bioavailability of toxic substances, such as mercury, with respect to contaminant speciation, fate, transport, and persistence in the aquatic system (Hudson et al. 1994; Tsiros and Ambrose 1998).

The fate of chemicals in the aquatic environment is determined by their reactivity and the rate of their physical transport through the environment. Knowledge of the chemical, biological, and physical reactions is needed to assess the speciation and partitioning mechanisms by which a contaminant is distributed in the aquatic system. These reactions can be better understood by applying the concept of the conservation of mass to perform quantitative analyses that are based on mass balance calculations which account for mass inputs, outputs, reactions, and accumulation, as described by the following equation (Chapra 1997):

$$\text{accumulation of mass} = \text{loadings} \pm \text{transport} \pm \text{reactions}$$

Here, *mass* refers to the mass of a particular constituent being modeled within a specified control volume, and *transport* represents the movement of mass through the control volume, equaling mass inputs minus mass outflows, together with water. In addition to this flow, mass can be gained or lost through reactions of constituents within the control volume. Reactions can also add mass to the system by changing a specific constituent into another that is being modeled. Finally, the mass of substance being modeled can be increased by loadings from external sources such as flow from outside the control volume. If the inputs are greater than the outflows during the period of calculation, the mass of the substance increases. On the other hand, if the outflows are greater than the inputs, the mass decreases. If there is no change in concentration with time, the system will be at steady state or dynamic equilibrium.

In addition to considerations of contaminant transport in aquatic systems, in recent years an assessment of the effects of chemical pollutants and the prediction of future chemical concentrations were added to considerations of contaminant transport in aqueous systems, under various loading scenarios or management action alternatives. As a result modelers are keen to develop kinetic models to determine chemical speciation. The intention is to identify equilibrium chemistry mechanisms and integrate them into the mass balance. Equilibrium models allow calculation of the speciation of elements such as mercury, which is important for determining not only their fate and transport but also their toxicity in aquatic ecosystems.

B. The Role of Data Quality in Modeling Mercury

During the last few years, more accurate site-specific studies have provided information for the development of models for mercury cycling and fate in aquatic systems, in lakes in particular (Hudson et al. 1994; Munthe 1994; Henry et al. 1995a,b; Leonard et al. 1995; Rudd 1995; Saouter et al. 1995; Vandal et al. 1995; Gbondo-Tugbawa and Driscoll 1998; Tsiros and Ambrose 1998). Improvements in sampling and analytical procedures have led to a general improvement in the knowledge of mercury chemistry and its biological transformations. Analytical techniques have lowered the limit of detection (Fitzgerald and Gill 1979; Gill and Fitzgerald 1987; Bloom and Fitzgerald 1988) from approximately 10^{-5} (10 mg L^{-1}) to 10^{-12} (1.0 ng L^{-1}) (Schnoor 1996). Consequently, most historical data for environmental mercury concentrations in water before the early 1980s should be viewed with caution, with the probability of bias toward inflated values resulting from sample contamination through various means (Wren 1995).

Determining the degree of reliability of analytical results is an essential step in the calculation of contamination levels and in the pursuance of changes of these levels with time and location. It is therefore of major importance in the development of improved models. Unreliable data are often a result of contamination problems during sample collection, handling, storage, or of analytical procedures that render the results questionable, especially those for mercury in aquatic systems. In an attempt to circumvent these problems Gill and Fitzgerald (1987) described an integrated analytical methodology, from sampling to analysis, that has been called ultraclean procedures. These procedures have improved sample collection and handling techniques by eliminating direct contact between the sampling equipment and field personnel, and also by employing clean laboratory conditions and practices.

Krabbenhoft and Babiarz (1992) pointed out that some techniques for ultraclean trace metal sampling and analysis in the marine environment were adapted to mercury analysis and lake studies. They noted that some studies demonstrated that, in remote areas, mercury concentrations in surface waters were generally less than 10 ng L^{-1}. The estimates were two to three orders of magnitude less than the best estimates from previous studies because of improvements in sample collection techniques and increased analytical sensitivity.

C. Existing Models of Mercury Cycling

The challenge now is to improve the understanding of mercury biogeochemistry to explain the series of processes that contribute to toxic or near-toxic concentrations in aquatic environments. To this end hydrodynamic models have been linked into biological and chemical models, which enables predictions of mercury concentration in water, sediment, and trophic levels.

At this point it is important to provide a brief description of a model that has been used to predict mercury concentration and bioaccumulation in lakes. It was developed by Hudson et al. (1994) as part of the Mercury in Temperate Lakes (MTL) study in Wisconsin, U.S. The mercury cycling model (MCM) is a deterministic simulation model that incorporates the major processes in the transport of mercury through boundaries in lakes. It considers atmospheric deposition, gas exchange, inflow and outflow of water, burial in sediments or scavenging by particles, chemical/biological transformations, and accumulation in biota.

The MCM employs a simple compartmental structure to represent the different geochemical environments and trophic levels in lake ecosystems (Fig. 3). The lake is divided into epilimnetic, hypolymnetic, and sedimentary compartments, with four trophic levels, i.e., phytoplankton, zooplankton, planktivorous fish, and piscivorous fish, representing the food chain. Ecosystem dynamics are simulated to obtain coherent productivity and trophic transfer fluxes for modeling bioaccumulation and biomagnification. The boundaries of the lake consist of the atmosphere, lake margins, and permanent sediments. The user must specify the data necessary to calculate input fluxes from the atmosphere, namely, mean atmospheric concentrations of Hg^0, Hg^{2+}, CH_3Hg^+, and particulate Hg^{2+}, and via water transport into the lake, i.e., concentrations of each mercury species in groundwater and tributaries. Biogeochemical reactions within the watershed are not considered, and burial in permanent sediments represents a one-way flux out of the surficial sediments.

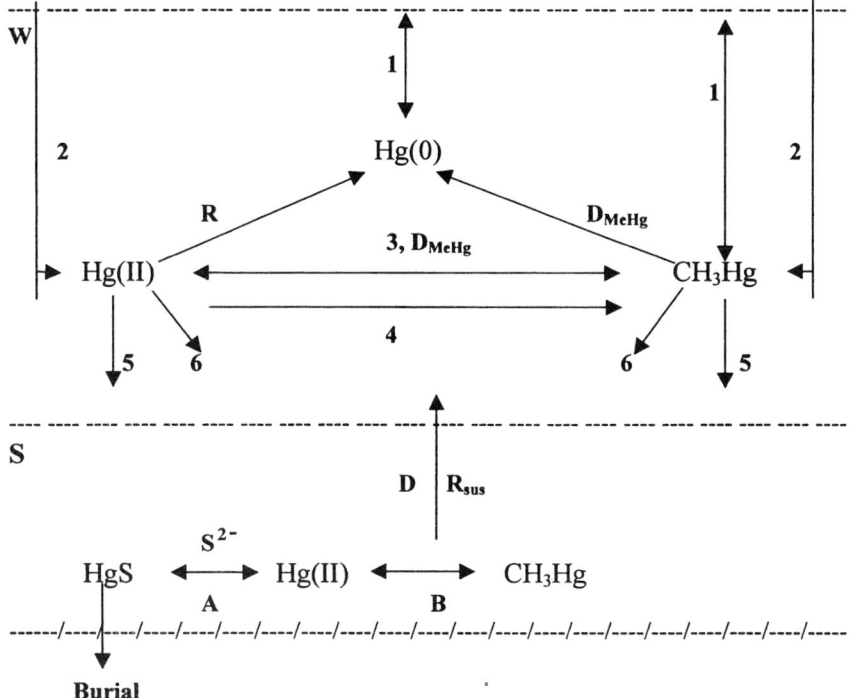

Fig. 3. The mercury cycling model (adapted from Hudson et al. 1994). 1, votilization and deposition; 2, deposition and runoff; 3, methylation; 4, biomagnification (phytoplankton → zooplankton → planktivorous fish → piscivorous fish); 5, sedimentation; 6, outflow; A, abiotic; B, bacteria; D, diffusion; D_{MeHg}, demethylation; R, reduction; R_{sus}, resuspension; S, sediment; W, water (epilimnion and hypolimnion).

The original dynamic MCM was simplified to a set of steady state mass conservation and rate equations, and was applied to investigate the dependencies of the rates of transport and transformation in the mercury cycle in each of the main compartments in seepage lakes. These lakes were remote from point sources and mercury-containing geological strata. Thus, only atmospheric sources were considered to quantify mercury cycling and mass balances (Porcella 1994). The steady-state MCM was limited to three components: Hg^0, Hg^{2+}, and CH_3Hg^+ with only the essential reactions (reduction, methylation, and demethylation) and transport processes (deposition, gas exchange, and scavenging by particles). The following conclusions were reached as a result of the MTL study (Hudson et al. 1994):

- In-lake methylation was the major source of CH_3Hg^+, although gas exchange provided a source of methylmercury comparable to atmospheric deposition.
- Reduction of Hg^{2+} and subsequent volatilization of Hg^0 was a major sink of mercury in high pH lakes. Inhibition of this mechanism at low pH led to significant increases in the production of CH_3Hg^+.
- An increase in overall methylation/demethylation ratios at low pH, as well as stimulus of methylation by sulfate, may have contributed to the observed increases in methylmercury in low-pH lakes.
- Hg^{2+} and CH_3Hg^+ underwent a significant degree of reaction with DOC.
- A negative correlation was found between the lake pH and mercury concentration in fish; this may be explained by the increased aqueous concentrations of methylmercury at low pH.

As a result of the same study, Watras et al. (1994) concluded that the bioconcentration factor for methylmercury increased by threefold in each trophic level, approaching 10^7 in fish. It was suggested that fish contamination would rise in response to small increases in net rates of methylmercury production, recycling, or loading.

Two other specific U.S. studies apply the MCM to the lakes. Firstly results from a research project conducted in Lakes Erie and Superior, performed by Leonard et al. (1995) for regulatory purposes, demonstrated that the MCM provides a reasonable approach to predicting mercury concentrations. The MCM simulated the partitioning of mercury among three forms between all biotic and abiotic components. In Lake Superior, almost all mercury inputs (93%) originated from deposition, and 60% of these inputs was buried in the sediments. The annual accumulation in fish was estimated to be 0.47 mg kg^{-1} carbon of fish. In Lake Erie, atmospheric deposition accounted for about 20% of all mercury inputs, whereas tributary inputs accounted for 80%. Like Lake Superior, most of the mercury (55%) in Lake Erie was partitioned to the sediments. However, the annual accumulation in fish was relatively small and was estimated to be 0.026 mg kg^{-1} carbon of fish. In both lakes methylmercury was the main species found in fish.

The application of the MCM to the Great Lakes did present some limitations, primarily because the epilimnion and hypolimnion were each considered to be completely mixed, and also because site-specific habitats could not be treated separately. According to the authors, other limitations included the lack of fish biomass estimates and sparse data for aqueous mercury concentrations.

Secondly, Gbondo-Tugbawa and Driscoll (1998) used a variant of the MCM, the regional mercury cycling model (R-MCM; version 1.0bBeta), to simulate the mass balance of total and methylmercury in Onondaga Lake in New York State, U.S. This enable them to evaluate possible remediation scenarios. The Onondaga Lake is an urban lake, considered highly contaminated because it has historically received loadings of industrial wastewater. High concentrations of total mercury and methylmercury in the water column, sediments, and fish have been found. The model predictions for concentrations of mercury and methylmercury in the water column and fish generally agreed with measured values. The authors concluded that outflow from the lake was the principal sink of total mercury, which was estimated to be 524.5 g m^{-2} yr^{-1} (77% of the total mercury sinks), and net methylation was the principal source for methylmercury, estimated to be 146.7 g m^{-2} yr^{-1} (84.8% of the total methylmercury inputs). It is worth noting that the model provided a reasonable estimate of in-lake methylmercury production if the rate of hypolimnetic accumulation of methylmercury estimated (0.23 g m^{-2}d^{-1}) is compared with the net methylmercury production estimated by the R-MCM (0.47 g m^{-2}d^{-1}). The principal sinks for methylmercury in the lake were net sedimentation (132.6 g m^{-2} yr^{-1}, 80% of the total methylmercury sink) and outflow (32.3 g m^{-2} yr^{-1}; 20% of the total methylmercury sink).

Henry et al. (1995a) presented results for the same lake, showing that the principal sink for total mercury was net sedimentation (936.7 g m^{-2}yr^{-1}) instead of outflow from the lake. However, the results for methylmercury agreed with the study by Gbondo-Tugbawa and Driscoll (1998). It was also concluded that net methylation was the principal source (52.5 g m^{-2} yr^{-1}), whereas sediment burial was the principal sink (50.3 g m^{-2} yr^{-1}). Model simulations showed that inputs of total mercury originated from tributary and wastewater inflow to the Onondaga Lake. In contrast, methylmercury was derived from internal production.

Studies have been carried out in Canada, as part of the Experimental Lakes Area Reservoir Project (ELARP), in an attempt to improve the understanding of production, storage, and export of methylmercury in wetlands. Rudd (1995) produced a model to predict mercury concentrations in fish in reservoirs, based on current knowledge that wetlands can act as external inputs of methylmercury to aquatic systems (Gilmour and Henry 1991). The author used data from studies conducted in Sweden (a drainage lake), Canada (flooded and unflooded boreal catchments), and the U.S. (a seepage lake) to calibrate the model. Modeling of data demonstrated that under different circumstances each of the three sources of mercury (internal production, inputs from watersheds that contain wetlands, and atmospheric deposition) could have significant effect on lakes. The model indicated that internal production of methylmercury is of particular significance for reservoirs.

However, while these models seem to work perfectly, they do not cover rivers. A modeling approach for rivers, discussed by Schnoor (1996), considers coupling a chemical equilibrium model with mass-balance equations for total metal concentration in the water column. This approach assumes that chemical equilibrium among metal species in solution and between phases is valid. It is also assumed that adsorption and desorption processes are fast relative to transport processes, and that suspended solids concentration is constant within a river segment. These assumptions allow the model to reduce the number of mass-balance equations from six (dissolved metal concentration, adsorbed metal concentration, and suspended solids concentration in the overlying water column and sediment) to only two equations (total metal concentration in the unfiltered water and adsorbed metal concentration in the sediment). Consequently, a mass-balance equation for every chemical species does not have to be formulated.

If chemical equilibrium does not apply, each chemical species of concern must be simulated with its own mass-balance equation. An example where chemical equilibrium does not apply would be the methylation and demethylation of mercury and elemental mercury volatilization, which are relatively slow processes. Kinetics controls bacterial methylation, demethylation, and bioconcentration of mercury in aquatic environments, and therefore an equilibrium approach is not sufficient. Thus, chemical equilibrium expressions must be coupled with mass-balance kinetics equations and solved simultaneously to find solutions to such problems. The model proposed by Schnoor (1996) to assess mercury contamination in rivers seems to be well defined, but no results were available to demonstrate the power of this model.

Although these models differ in their databases, input–output formats, methods of handling heterogeneous equilibrium, and programming codes, they have been developed for the same general purpose, i.e., to predict mercury concentrations in aqueous media.

D. Research and Modeling Requirements Concerning Mercury in Aquatic Systems

Chemical modeling can be used to describe the chemical characteristics of an aquatic environment. When the requisite input data for the system to be modeled, as well as adequate reference data (thermodynamic and kinetic), are given (Jenne 1979), lakes and wetlands have been privileged while information is scarce for rivers. The lack of knowledge of some specific aquatic environments and the scarcity of the available data seem to be the major obstacles to presenting a properly constructed model.

The primary objectives for future investigations should be to determine the environmental variables and processes that influence mercury bioavailability and methylmercury bioaccumulation in the aquatic food chains, to establish the lowest effect level for human exposure to methylmercury, to quantify mercury sources and pools across a variety of ecosystems to incorporate advances into simulation models, and to test and apply the models to various environmental problems. Suggested areas of site-specific research include sites and seasonality of methylation rates in wetlands, with special attention to both biotic and abiotic processes; interactions between mercury and nutrients, sulfur, and other metals; resolving local/regional variability in mercury deposition; and the importance of reduction of Hg^{2+} to Hg^0 (Hartung 1994b; Zillioux et al. 1993; Fitzgerald et al. 1994). Additionally, current research seeks to identify the organisms that mediate the demethylation processes, to quantify where and under what conditions each process dominates, and to determine rates that govern the reactions (Krabbenhoft 1996a).

It is also necessary to quantify fluxes of methylmercury under different environmental circumstances. In his 1995 study, Rudd listed a number of studies that should be undertaken to (a) improve knowledge of total rates of atmospheric deposition of methylmercury at different geographic locations; (b) determine if methylmercury export from wetlands is higher in regions of high atmospheric deposition than regions of low atmospheric deposition; (c) develop methods that accurately determine rates of methylation in lakes and wetlands; and (d) develop methods that accurately determine rates of mercury demethylation in uplands, wetlands, and in lakes. If bioaccumulation of mercury in remote lakes is to be understood, it is necessary to investigate the environmental dependencies of factors that regulate the formation, destruction, and trophic transfer of methylmercury.

V. Conclusions

Although mercury may only be present at very low concentrations in the water column, because of a potentially high degree of bioaccumulation such concentrations remain a cause for concern, particularly where fish are caught and consumed locally. A centrally important phenomenon controlling the toxicological impact of mercury in the environment is the methylation process. However, it is becoming apparent that there may be distinct methylation processes, which suggests several important factors may be influencing the transformation processes.

To date, the models used to identify and predict mercury concentrations in aquatic systems were developed for specific ecosystems with uniquely individual characteristics. A case in point is the model that was specifically set up for the Florida Everglades physical system, a complex landscape of marshes, canals, impoundments, and agricultural fields, with a total area of 1,036,000 ha (4,000 square miles) (Tsiros and Ambrose 1998). Due to the peculiarity of the system, where mercury is a significant issue, this model is not applicable to other environments (David Krabbenhoft, personal communication).

Despite the efforts to identify and quantify the major factors responsible for mercury speciation and dispersion in the food chain, many of the mechanisms that govern mercury partitioning and fate are still poorly understood. Metabolism, concentrating ability, absorption, and excretion rates remain undefined and unquantified. Up-coming studies should be focused on research into ways of applying the results determined by modeling a set of data for one site to another during the confirmation step of the model. Thus, the rate constants and partitioning constants could be kept unchanged for different sites with only a change in the site conditions.

Summary

In response to increasing scientific evidence of the environmental toxicity of mercury and its organic compounds, this study reviews the state of knowledge about the mercury cycle in aquatic systems. It describes the aquatic chemistry of mercury and discusses the importance of biological and physicochemical parameters such as pH, temperature, dissolved organic carbon, oxygen concentration, mercury and methylmercury concentration and availability, as well as sulfate, manganese, and iron concentration in surfaced waters. There is still a paucity of information on environmental dependency of factors that regulate the formation, destruction, and trophic transfer of methylmercury. This lack has led to numerous studies to define the factors that can influence its bioconcentration and bioaccumulation. This review presents some alternative models for mercury cycling, speciation, and partitioning based on the trend toward "site-specific water quality standards," in which chemical speciation is considered on a site-by-site basis.

References

Akagi H, Malm O, Kinjo Y, Harada M, Branches FJP, Pfeiffer WC, Kato H (1995) Methylmercury pollution in the Amazon, Brazil. Sci Total Environ 175: 85–95.
Allan CJ, Heyes A (1998) A preliminary assessment of wet deposition and episodic transport of total and methylmercury from low order blue ridge watersheds, SE, USA. Water Air Soil Pollut 105:573–592.
Aula I, Braunschweiler H, Malin I (1995) The watershed flux of mercury examined with indicators in the Tucurui reservoir in Para, Brazil. Sci Total Environ :97–107.
Bailey RA (1978) The environmental chemistry of some important elements. In: Bailey RA (ed) Chemistry of the environment. Academic Press, New York, pp 392–394.
Balogh SJ, Meyer ML, Johnson DK (1998) Transport of mercury in three contrasting river basins. Environ Sci Technol 32(4):456–462.
Bigham GN, Vandal GM (1996) A drainage basin perspective of mercury transport and bioaccumulation: Onondaga Lake, New York. Neurotoxicology 17(1):279–290.
Bjornberg A, Hakanson L, Lundberg K (1988) A theory on the mechanisms regulating the bioavailability of mercury in natural waters. Environ Pollut 49:53–61.
Bloom N, Fitzgerald WF (1988) Determination of volatile mercury species at the picogram level by low-temperature gas chromatography with cold-vapour atomic fluorescence detection. Anal Chim Acta 208:151–161.
Brosset C (1981) The mercury cycle. Water Air Soil Pollut 16:253–255.
Bubb JM, Kirk PWW, Beck MB, Wheater HS, Lester JN (1988) Mercury in the River Yare and its associated broad: survey and modelling. In: Astruc M, Lester JN (eds) Proceedings of the international conference on heavy metals in the hydrological cycle. Selper, UK, 137–154.

Bubb JM, Rudd T, Lester JN (1991) Distribution of heavy metals in the River Yare and its associated broad: mercury and methylmercury. Sci Total Environ 102:47–168.

Chapra SC (1997) Surface Water Modelling. McGraw-Hill, New York.

Cossa D, Mason RP, Fitzgerald WF (1994) Chemical speciation of mercury in a meromitic lake. In: Watras CJ, Huckabee JW (eds) Mercury pollution: integration and synthesis. Lewis, Chelsea, MI, pp 57–67.

Doi RN, Ui J (1973) The distribution of mercury in fish and its form of occurrence. In: Krenkel PA (ed) Heavy metals in the aquatic environment: an international conference. Pergamon Press, New York, pp 197–227.

Downs SG, Macleod CL, Lester JN (1998) Mercury precipitation and its relation to bioaccumulation in fish: a literature review. Water Air Soil Pollut 108:149–187.

Fargerström T, Jernelov A (1972) Some aspects of the quantitative ecology of mercury. Water Res 6:1193–1202.

Fitzgerald WF, Gill GA (1979) Subnanogram determination of mercury by two-stage gold amalgamation and gas phase detection applied to atmospheric analysis. Anal Chem 51(11):1714–.

Fitzgerald WF, Mason RP, Vandal GM, Dulac F (1994) Air-water cycling of mercury in lakes. In: Watras CJ, Huckabee JW (eds) Mercury pollution: integration and synthesis. Lewis, Chelsea, MI, pp 203–220.

Fleisher M (1970) Summary of the literature on the inorganic geochemistry of mercury. Mercury in the environment. US Geol Surv Prof Pap 713:6–13.

Gavis J, Ferguson JF (1972) The cycling of mercury through the environment. Water Res 6:989–1008.

Gbondo-Tugbawa S, Driscoll CT (1998) Application of the regional mercury cycling model (RMCM) to predict the fate and remediation of mercury in Onondaga Lake, New York. Water Air Soil Pollut 105:417–426.

Gill GA, Fitzgerald WF (1987) Picomolar mercury measurements in seawater and other materials using stannous chloride reduction and two-phase gold amalgamation with gas phase. Mar Chem 20:227–243.

Gilmour CG, Henry EA (1991) Mercury methylation in aquatic systems affected by acid deposition. Environ Pollut 71:131–163.

Greeson PE (1970) Biological factors in the chemistry of mercury. Mercury in the environment. US Geol Surv Prof Pap 713:32–34.

Guimarães JRD, Malm O, Pfeiffer WC (1995) A simplified radiochemical technique for measurements of net mercury methylation rates in aquatic systems near gold mining areas, Amazon, Brazil. Sci Total Environ 175:151–162.

Hartung R (1994a) The role of food chains in environmental mercury contamination. In: Hartung R, Dinman DB (eds) Environmental Mercury Contamination, Part III: Environmental dynamics of mercury. Ann Arbor Science, Ann Arbor, MI, pp 172–174.

Hartung R (1994b) Research needs: study of the environmental dynamics of mercury. In: Hartung R, Dinman DB (eds) Environmental Mercury Contamination, Part III: Environmental dynamics of mercury. Ann Arbor Science, Ann Arbor, MI, pp 197–198.

Hem JD (1970) Chemical behaviour of mercury in aqueous media. Mercury in the environment. US Geol Surv Prof Pap 713:19–24.

Henry EA, Dodge-Murphy LJ, Bigham GN, Klein SM (1995a) Total mercury and methylmercury mass balance in an alkaline, hypereutrophic urban lake (Onondaga Lake, NY). Water Air Soil Pollut 80:509–518.

Henry EA, Dodge-Murphy LJ, Bigham GN, Klein SM, Gilmour CC (1995) Modelling the transport and fate of mercury in an urban lake (Onondaga Lake, NY). Water Air Soil Pollut 80:489–498.

Hudson JM, Gherini SA, Watras CJ, Porcella D (1994) Modelling the biogeochemical cycle of mercury in lakes: the mercury cycling model (MCM) and its application to the MTL study lakes. In: Watras CJ, Huckabee JW (eds) Mercury pollution: integration and synthesis. Lewis, Chelsea, MI, pp 473–523.

Hultberg H, Iverfelt A, Lee YH (1994) Methylmercury input/output and accumulation in forested catchments and critical loads for lakes in southwestern Sweden. In: Watras CJ, Huckabee JW (eds) Mercury pollution: integration and synthesis. Lewis, Chelsea, MI, pp 313–321.

Hurley JP, Watras CJ, Bloom NS (1994) Distribution and flux of particulate mercury in four stratified seepage lakes. In: Watras CJ, Huckabee JW (eds) Mercury pollution: integration and synthesis. Lewis, Chelsea, MI, pp 69–82.

Jacobs LA, Klein SM, Henry EA (1995) Mercury cycling in the water of a seasonally anoxic urban lake (Onondaga Lake, NY). Water Air Soil Pollut 80:553–562.

Jenne EA (1970) Atmospheric and fluvial transport of mercury. Mercury in the environment. US Geol Surv Prof Pap 713:40–49.

Jenne EA (1979) Chemical modelling: goals, problems, approaches and properties. In: Jenne EA (ed) Chemical modelling in aqueous systems. ACS Symp Series 3–21.

Jensen S, Jernelöv A (1972) Behaviour of mercury in the environment. Mercury contamination in man and his environment. IAEA Techn Rep Ser 137:43–47.

Jernelöv A (1994b) Mercury and food chains. In: Hartung R, Dinman DB (eds) Environmental Mercury Contamination, Part III: Environmental Dynamics of Mercury. Ann Arbor Science, Ann Arbor, MI, pp 174–177.

Jernelöv A, Åsell B (1973) The feasibility of restoring mercury-contaminated waters. Heavy metals in the aquatic environment: an international conference. Pergamon Press, New York, pp 299–309.

Jernelöv, A (1994a) Factors of transformation of mercury to methylmercury. In: Hartung R, Dinman DB (eds) Environmental Mercury Contamination, Part III: Environmental dynamics of mercury. Ann Arbor Science, Ann Arbor, MI, pp 167–172.

Johansson K, Iverfeldt A (1994) The relation between mercury content in soil and transport of mercury from small catchments in Sweden. In: Watras CJ, Huckabee JW (eds) Mercury pollution: integration and synthesis. Lewis, Chelsea, MI, pp 527–539.

Kelly CA, Rudd JWM, St. Louis VL, Heyes A (1995) Is total mercury concentration a good predictor of methylmercury concentration in aquatic systems? Water Air Soil Pollut 80:714–724.

Krabbenhoft DP (1996a) Summary document for the U.S. Geological Survey workshop on mercury cycling in the environment. U.S. Geological Survey, Golden, Colorado, July 7–9, 1996.

Krabbenhoft DP (1996b) Mercury studies in the Florida Everglades. U.S. Geological Survey Fact Sheet FS-166-96, U.S. Department of the Interior, Washington, DC.

Krabbenhoft DP, Babiarz CL (1992) The role of groundwater in aquatic mercury cycling. Water Res 28(12):3119–3128.

Krabbenhoft DP, Benoit JM, Babiarz CL, Hurley JP, Andren AW (1995) Mercury cycling in the Allequash creek watershed, northern Wisconsin. Water Air Soil Pollut 80(1–4): 425–433.

Krabbenhoft DP, Rickert DA (1995) Mercury contamination of aquatic ecosystems. U.S. Geological Survey Fact Sheet FS 216-95, U.S.G., Washington, DC.

Leonard D, Reash R, Porcella D, Pakalkar A, Summers K, Gherini S (1995) Use of the mercury cycling model (MCM) to predict the fate of mercury in the Great Lakes. Water Air Soil Pollut 80:519–528.

Lindqvist O (1994) Atmospheric cycling of mercury: an overview. In: Watras CJ, Huckabee JW (eds) Mercury Pollution: Integration and Synthesis. Lewis, Chelsea, MI, pp 181–185.

Lindqvist O, Rodhe H (1985) Atmospheric mercury: a review. Tellus 37B:136–159.

Malm O, Castro MB, Bastos WR, Branches FJP, Guimarães JRD, Zuffo CE, Pfeiffer WC (1995) An assessment of Hg pollution in different gold mining areas, Amazon Brazil. Sci Total Environ 175:127–140.

Mattice JS, Porcella DB, Broksen RW (1997) Sediment-water interactions affect assessments of metals discharges at electrical utilities. Water Air Soil Pollut 99:187–199.

Meili M (1988) Bioaccumulation of mercury in aquatic ecosystems: a biological approach. In: Heavy Metals in the Hydrological Cycle, pp 249–256.

Meili M (1994) Aqueous and biotic mercury concentrations in boreal lakes: model predictions and observations. In: Watras CJ, Huckabee JW (eds) Mercury Pollution: Integration and Synthesis. Lewis, Chelsea, MI, pp 99–106.

Meuleman C, Lemarks M, Baeyens W (1995) Mercury speciation in Lake Baikal. Water Air Soil Pollut 80:539–551.

Morel FMM, Hering JG (1993) Principles and Applications of Aquatic Chemistry. Wiley, New York.

Morrison KA, Therrien N (1994) Mercury release and transformation from flooded vegetation and soils: experimental evaluation and simulation modelling. In: Watras CJ, Huckabee JW (eds) Mercury Pollution: Integration and Synthesis. Lewis, Chelsea, MI, pp 355–365.

Munthe J (1994) The atmospheric chemistry of mercury: kinetic studies of redox reactions. In: Watras CJ, Huckabee JW (eds) Mercury Pollution: Integration and Synthesis. Lewis, Chelsea, MI, pp 273–279.

Porcella DB (1994) Mercury in the environment: biogeochemistry. In: Watras CJ, Huckabee JW (eds) Mercury Pollution: Integration and Synthesis. Lewis, Chelsea, MI, pp 3–19.

Porvari P (1995) Mercury levels of fish in Tucurui hydroelectric reservoir and in the River Moju in Amazonia, in the state of Para, Brazil. Sci Total Environ 175:109–117.

Rudd JWM (1995) Sources of methylmercury to freshwater to ecosystems: a review. Water Air Soil Pollut 80:697–713.

Saouter E, Gillman M, Turner R, Barkay T (1995) Development and field validation of microcosm to simulate the mercury cycle in a contaminated pond. Environ Toxicol Chem 14:69–77.

Scherbatskoy T, Shanley JB, Keeler GJ (1998) Factors controlling mercury transport in an upland forested catchment. Water Air Soil Pollut 105:427–438.

Schnoor JL (1996) Modeling Trace Metals. In: Schnoor JL, Zenhder A (eds) Environmental Modeling: fate and transport of pollutants in water, air, and soil. Wiley Interscience, New York, pp 331–354.

Schroeder WH (1988) Mercury species in the biological cycle. In: Astruc M, Lester JN (eds) Heavy Metals in the Hydrological Cycle. London, pp 83–90.

Stumm W, Morgan JJ (1995) Aquatic Chemistry: Chemical Equilibria and Rates in Natural Waters. Wiley, New York.

Taylor SR (1964) Abundance of chemical elements in the continental crust: a new table. Geochim Cosmochim Acta 28:1273–1285.

Tsiros IX, Ambrose RB (1998) Environmental screening modeling of mercury in the upper Everglades of South Florida. J Environ Sci Health A33(4):497–525.

USEPA (1990) Technical guidance manual for performing waste load allocations. Book III: Estuaries. Part 1: Estuaries and waste load allocation models. Supplement V: Metals. U.S. Environmental Protection Agency, Washington, DC, pp 2–35.

Vandal GM, Fitzgerald WF, Rolphus KR, Lamborg CH (1995) Modelling the elemental mercury cycle in Pallete Lake, Wisconsin, USA. Water Air Soil Pollut 80:529–538.

Verta M; Matelsinen T; Porveri P; Nieni M; Uuri-Rauve A; and Bloom NS (1994) Methylmercury sources in boreal lake ecosystems. In: Waters CY, Huckabee JW (eds) Mercury Pollution: Integration and Synthesis. Lewis, Chelsea, MI, pp. 119–136.

Wang JS, Huang PM, Hammer UT, Liaw WK (1989) Role of dissolved oxygen in the desorption of mercury from freshwater sediment. In: Nriagu JS, Lakshmirarayana JSS (eds) Aquatic Toxicology and Water Quality Management, Vol. 22. Wiley, New York, pp 153–159.

Watras CJ, Bloom NS (1994) The vertical distribution of mercury species in Wisconsin lakes: accumulation in plankton layers. In: Watras CJ, Huckabee JW (eds) Mercury Pollution: Integration and Synthesis. Lewis, Chelsea, MI, pp 137–152.

Watras CJ, Bloom NS, Hudson RJM, Gherini S, Munson R, Claas SA, Morrison KA, Hurley J, Wiener JG, Fitzgerald WF, Mason R, Vandal G, Powell D, Rada R, Rislov L, Winfrey M, Elder J, Krabbenhoft D, Andren AW, Babiarz C, Porcella D, Huckabee JW (1994) Sources and fates of mercury and methylmercury in Wisconsin Lakes. In: Watras CJ, Huckabee JW (eds) Mercury Pollution: Integration and Synthesis. Lewis, Chelsea, MI, pp 153–177.

Wood JM (1973) Metabolic cycles for toxic elements in the environment: a study of kinetics and mechanism. In: Krenkel PA (ed) Heavy Metals in the Aquatic Environment: An International Conference. Pergamon Press, New York, pp 105–115.

Wood JM, Penley MW, DeSimone KE (1972) Mechanisms for methylation of mercury in the environment. Mercury contamination in man and his environment. IAEA Tech Rep Ser 137. IAEA, Geneva, pp 43–47.

World Health Organization WHO (1990) Methylmercury. Environmental Health Criteria 101. Geneva.

World Health Organization WHO (1991) Inorganic Mercury. Environmental Health Criteria 118. Geneva.

Wren CD, Harris S, Hattrup N (1995) Ecotoxicology of mercury and cadmium. In: Hoffman K (ed) Handbook of Ecotoxicology. Lewis, Chelsea, MI, pp 392–423.

Zillioux EJ, Porcella DB, Benoit JM (1993) Mercury cycling and effects in freshwater wetland ecosystems. Environ Toxicicol Chem 12:2245–2264.

Manuscript received May 27, 1999; accepted June 2, 1999

Biomarkers in Terrestrial Invertebrates For Ecotoxicological Soil Risk Assessment

Jan E. Kammenga·Reinhard Dallinger·Marianne H. Donker·
Heinz-R. Köhler·Vibeke Simonsen·Rita Triebskorn·Jason M. Weeks

Contents

I. Introduction	94
II. Stress Proteins	97
III. Metallothioneins and Other Metal-Binding Proteins	104
IV. Histology and Ultrastructure	109
V. Isozymes	113
VI. Lysosomal Membrane Integrity	115
VII. Novel Biomarkers	118
VIII. Confounding Factors and Transiency of the Biomarker Response	120
IX. Ecological Relevance of Biomarkers in Soil Invertebrates	125
X. Biomonitoring Using Soil Invertebrate Biomarkers	126
A. Chemical Residue Measurement	126
B. Choice of Taxa and Complementary Activities	126

Communicated by George W. Ware

J.E. Kammenga (✉)
Laboratory of Nematology, Wageningen University, Binnenhaven 10, 6709 PD Wageningen, The Netherlands

R. Dallinger
Institut für Zoologie Abteilung Zoophysiologie, Technikerstraße 25, Universität Innsbruck, A 6020 Innsbruck, Austria

M.H. Donker
Department of Ecology and Ecotoxicology, Vrije Universiteit, De Boelelaan 1087, 1081 HV Amsterdam, The Netherlands

H.-R. Köhler
Zoological Institute, Department of Cell Biology, University of Tübingen, Auf der Morgenstelle 28, D-72076 Tübingen, Germany

V. Simonsen
National Environmental Research Institute, Department of Terrestrial Ecology, PO Box 314, Vejlsovej 25, DK-8600 Silkeborg, Denmark

R. Triebskorn
Zoological Institute, Department of Physiological Ecology of Animals, University of Tübingen, Auf der Morgenstelle 28, D-72076 Tübingen, Germany

J.M. Weeks
Institute of Terrestrial Ecology, Monks Wood, Abbots Ripton, Huntingdon PE17 2LS, UK.

XI. Future Potential and Limitations for Soil Risk Assessment
    Using Biomarkers 129
    A. Biomarker Potential 129
    B. Biomarker Limitations 130
XII. Conclusions 131
Summary 134
Acknowledgments 135
References 135

I. Introduction

In recent years there has been an increasing interest in the use of biomarkers in terrestrial invertebrates for the assessment of the potential adverse effects of chemicals on soil ecosystems. Terrestrial invertebrates offer meaningful targets because they play a major role in the functioning of the soil ecosystem by enhancing soil structure and the decomposition of organic material. Furthermore, invertebrates represent a major component of all animal species in soils and often are present in high population densities; thus, samples can be taken for analysis without significantly affecting population dynamics. In addition, there are ethical and legal considerations favoring their use in contrast to the use of vertebrates. In connection to biomarkers, soil invertebrates have the advantage that they are in direct contact with soil pore water or food exposure, in contrast to many vertebrates that are indirectly exposed through the food chain.

Biomarkers in terrestrial invertebrates have been used by ecotoxicologists to document and quantify both exposure to and effects of environmental pollutants. As an indicator for exposure, the biomarker response in terrestrial invertebrates is directly related to the bioavailable fraction of pollutants in soils. Before we evaluate the potential and limitations of biomarkers in terrestrial invertebrates for soil risk assessment, we need to define the term biomarker more clearly. The word biomarker may be used in the broadest sense to include almost any response measurement reflecting an interaction between a biological system and a potential hazard, which may be chemical, physical, or biological. In this review, we restrict ourselves to a biomarker in the sense of a measurable indicator of a chemical hazard.

The type of biomarker responses that could be considered range from the molecular to effects on the intact organism, the population or community structure, and perhaps also the structure and function of ecosystems (Peakall 1994). Figure 1 shows the cascade of events leading to organism death and where biomarkers should be applied to predict negative effects on an ecosystem. If biomarkers are to be used for soil risk assessment they must be applied at the lower end of the continuum (as indicated) where responses are sensitive and rapid while being reasonably easy to interpret.

In line with the ecotoxicological literature, we focus the term biomarker further to the detection of molecular, biochemical, physiological, or cellular alterations in organisms following exposure to pollutants (Peakall and Shugart 1992;

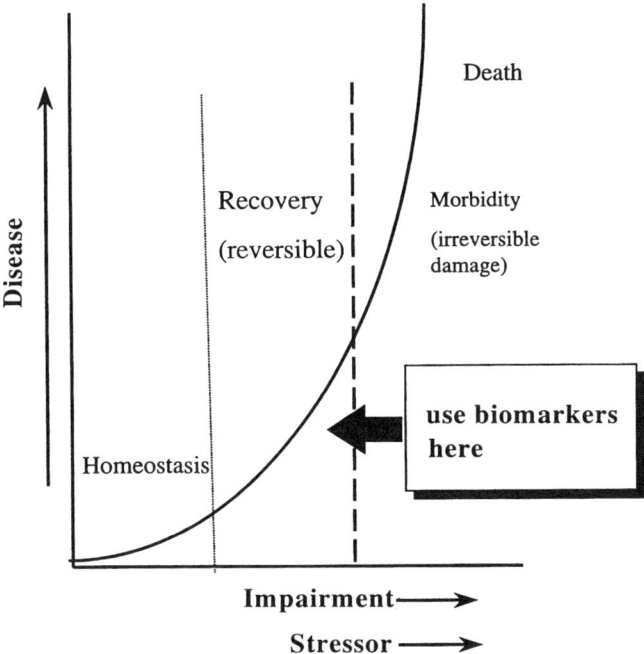

Fig. 1. The biomarker continuum (after Weeks et al. 1999).

Depledge and Fossi 1994). According to this definition, three classes of biomarkers can be identified:

1. Biomarker of exposure: response to an interaction between a xenobiotic agent and some target molecule or cell that is measured in a compartment within an organism
2. Biomarker of effect: an alteration that, depending upon the magnitude of the response, can be recognized as associated with an established or possible health impairment or disease
3. Biomarker of susceptibility: an indicator of an inherent or acquired ability of an organism to respond to the challenge of exposure to a specific xenobiotic substance

Following this division in classes it is important to note that there is an essential differerence between the use of biomarkers in aquatic (mainly freshwater streams) and terrestrial species. Aquatic biomarkers are often promoted for their potential to serve as an early warning of toxic stress in a fluctuating environment. Terrestrial biomarkers, however, are not primarily necessary for early warning because in soil ecosystems the half-life of toxic compounds is relatively long (often the soil serves as a sink for pollutants) and therefore the duration of exposure is also long. The surplus value of terrestrial biomarkers is their role in providing a better estimation of real "harm" to a soil system than measuring the total concentration of a contaminant in the soil compartment. The biomarker response may be a sensitive indicator of chemical stress before sublethal effects, such as inhibition of growth or reproduction, become apparent. Detection of biomarkers provides a powerful tool for the early assessment of exposure or effect of environmental contaminants at the below-individual level (Van Gestel and Van Brummelen 1996). Ideal characteristics of biomarkers should meet the following criteria:

1. Reflect the interaction (qualitative or quantitative) of the host biological system with the chemical of interest. There should be a clear and unambiguous concentration or dose–effect relationship between the concentration of the chemical in either the ambient environment or internal body burden and the biomarker response. Deviations from the targeted stress response should be distinguishable from other confounding stress factors.
2. Have a known and appropriate specificity and sensitivity to the interaction. Some biomarkers are highly specific for a particular compound or class of compounds, e.g., esterases and metal-binding proteins, while others such as heat shock proteins may be less specific.
3. Be reproducible qualitatively and quantitatively with respect to time, either short or long term. This point includes that the response transience should be known.
4. Be common to individuals within a population and to be of a defined variability within the normal, nonexposed population or group of interest.

The aim to protect soil ecosystems against chemical pollutants by the European Union (EU) has stimulated soil invertebrate biomarker research. The EU have funded two projects involving five member countries that were dedicated to the development and implementation of biochemical fingerprint techniques for assessing the exposure and effect of contaminants in soil invertebrates. The BIOPRINT project, which ran from 1994 to 1996, focused on the development of techniques for the detection of the following biomarkers (Kammenga 1995; 1996): (i) heat shock proteins, (ii) metallothioneins and metal-binding proteins, (iii) esterases, and (iv) lysosomal integrity. These biomarkers were selected because they have already been thoroughly studied as indicators for exposure and effect in mammals and aquatic organisms. The research program included organisms of different taxonomic and ecological groups that play a vital role in soil ecosystem processes, i.e., nematodes, diplopods, isopods, Collembola, gastropods, and oligochaetes. Nematodes, diplopods, Collembola, isopods, oligochaetes, and gastropods are key intermediaries in decomposition processes of organic matter in the soil. Furthermore, isopods and, to a lesser degree, also diplopods play an important role in the fragmentation of plant material, and oligochaetes enhance the soil structure by their burrowing activities. The quality of the soil depends, among others, on the optimal functioning of these soil organisms.

A second EU-funded project (BIOPRINT-II) was started for 3-yr in 1996. The main objective of this second project was the deployment of biochemical fingerprint techniques, which were developed in the BIOPRINT project (Kammenga and Simonsen 1997), for assessing the exposure and effect of toxicants in soil invertebrates in the field. The importance of this approach has been stressed by several authors (Peakall and Shugart 1992; Lagadic et al. 1994).

In this review we evaluate the future potential and current limitations of various biomarkers in soil invertebrates for the ecotoxicological risk assessment of soil pollution. We do not attempt to provide an exhaustive summation of the different biomarkers currently available but merely a critical review of the benefits and strengths of a set of biomarkers, including, those used within the BIOPRINT projects in view of their application to future soil risk assessment procedures. In Sections II to VI we review the history and current applications of selected biomarkers in terrestrial invertebrates, after which we discuss the advantages and problems involved in applying them using the criteria put forward.

Section VII concerns recent discoveries on the development of novel biomarkers in soil invertebrates such as histidine and nuclear magnetic resonance (NMR) profiling. Because biomarkers often respond to abiotic stress factors other than the presence of toxicants, Section VIII addresses the influence of confounding factors on the biomarker response to toxic insults. To be used in risk assessment procedures, the response transiency should be known, and this is also discussed in this section. The use of biomarkers in environmental risk assessment in relation to higher organisation levels such as the population or community level is discussed in Section IX. Finally, the value of biomarkers in terrestrial invertebrates for biomonitoring programs and risk assessment procedures is discussed in Sections X and XI, respectively.

II. Stress Proteins

Stress proteins constitute a set of protein families of different molecular weights that have long been called "heat shock proteins" (hsp) (Gething and Sambrook 1992). Originally discovered in *Drosophila* in response to increased temperature (Tissiéres et al. 1974), the induction of stress proteins has been found to occur ubiquitously after exposure to a variety of chemicals and also in response to injury, viral infections, and damage to tissues (summarized in Schlesinger et al. 1982, 1990; Nover 1984). Despite this apparent nonspecificity, the suitability of the stress response as a biomarker of exposure and effect has received increased interest in the last years and has been discussed recently (Peakall and Walker 1994; Sanders and Dyer 1994; De Pomerai 1996). Based on their molecular weight, stress proteins can be classified into different protein families. A heterogeneous group of low molecular weight stress proteins (LMW) with a molecular weight (MW) of about 15–40 kDa, a group of mitochondrial or cytoplasmic stress-60 (hsp60, chaperonin, cpn60, TCP-1) proteins (58–60 kDa), the prominent family of stress-70 (hsp70, BiP, 66–78 kDa), and the groups of stress-90 (hsp90, 83–90 kDa), and high molecular weight stress proteins (HMW; 100–110 kDa) all exist in invertebrates (Sanders 1993). Additionally, a small protein of 7 kDa, ubiquitin, which is involved in nonlysosomal protein degradation and another protein of ~10 kDa associated with stress-60 usually are assigned to the stress protein family.

The physiological legitimation to use stress proteins as biomarkers of exposure to or effect of toxicants is based on the mode of induction of at least the stress-70 family. Because stress-70 is involved in intracellular protein folding and membrane translocation (Pelham 1986; Morimoto et al. 1990), it is necessary for this stress protein family to bind uncoiled and partially folded polypeptide chains. Binding to polypeptides competitively releases an originally stress-70-associated protein of the heat shock factor (hsf) protein class, which in turn binds as a phosphorylated trimer to the heat shock element (hse) region of the DNA, and subsequently facilitates the expression of stress-70 itself (summarized by Gething and Sambrook 1992). Thus, whenever a stressor leads to an accumulation of denatured or malfolded protein in the cell, the elevated presence of unfolded segments or loops projecting from those proteins leads to an enhanced running-down of the aforementioned cascade of stress-70 induction ("abnormal protein hypothesis"; Edington et al. 1989; Sorger and Nelson 1989; Craig and Gross 1991). Chaperonin is also involved in the correct folding of proteins, but the exact mechanism of induction of this and all the other stress proteins remains unclear at the present. The mitochondrial hsp60 was shown by Martin et al. (1992) to form complexes with various polypeptides in organelles exposed to heat stress. They suggested a general mechanism in which hsp60 binding to a native reductase in the course of denaturation prevented its aggregation and restored the refolding at increased temperatures.

Based on this background, it is now well accepted that the main advantage of stress proteins as biomarkers is their ability to integrate effectively overall adverse effects on protein integrity, which are collectively summarized by the term "proteotoxicity" (Sanders 1993). On the other hand, an increased stress protein level generally, indicates the presence of a stressor, but cannot give any information on its (chemical) nature (Peakall and Walker 1994). Many of the numerous members of the aforementioned stress protein families, however, are regulated in a stressor-specific manner and this fact provides the potential to distinguish between at least different classes of stressors (Sanders and Dyer 1994).

In comparison to the presence of numerous papers on the stress protein response in vertebrates or vertebrate cell lines, information on stress proteins in terrestrial invertebrates is, except for the preferred objects of molecular biology, *Drosophila* spp. and *Caenorhabditis elegans*, rather scarce and mostly restricted to pathogenic animal parasites. The use of stress proteins as biomarkers for environmental hazards in invertebrates has received increased attention during the past few years only, and research in this field has predominantly focused on the aquatic environment (reviewed by Sanders 1993). Over the past few years, however, interest on the induction of stress protein expression by toxicants in terrestrial invertebrates has grown, and current approaches to establish biotests have led to promising results. The potential of various stress proteins as suitable biomarkers of exposure and effect is now reviewed.

Stress-90 (hsp90). Although stress-90 has been shown to bind the aryl hydrocarbon (Ah) receptor (Perdew 1988) and thus has a close connection to the cytochrome P-450 biotransformation system, this stress protein group has not been subjected to many ecotoxicological studies. Reviewing several papers on the cellular role of stress-90 led Sanders (1993) to summarize that upon exposure to environmentally stressful conditions the synthesis of stress-90 was increased. Indeed, Amaral et al. (1988) have shown that a 92-kDa stress protein was induced by arsenite in the protozoan *Tetrahymena pyriformis*. The specific mechanisms involved in this induction, however, have not been identified (Sanders 1993), and also the suitability of stress-90 as a biomarker remains unclear. In *Drosophila* the equivalent of hsp 90 is hsp 83 (chromosomal site 63bc). In *Drosophila* cells, hsp 83 interacts with gene expression of steroid hormones (Plesofsky-Vig 1996). Of hsp90 it is known that hsp90 acts only on specific protein targets (steroid hormone receptors, kinases, and calmodulin) all involved in signal transduction (Craig et al. 1994); therefore, it can be assumed that hsp90/83 is a specific biomarker for toxicants interacting with hormone (steroid) receptors.

Stress-70 (hsp70). Among the different stress protein families, the stress-70 group is undoubtedly the best studied, and together with stress-60 much research has focused on the establishment of a biotest carried out on this protein family. In laboratory tests, the increased expression of a set of stress-70 proteins after exposure to arsenite was shown for a permanent *Drosophila* cell line (Vincent and Tanguay 1982) and for *T. pyriformis* (Amaral et al. 1988). The first indications of the potential use of the stress-70 induction in terrestrial invertebrates for toxicity assessment not only in the laboratory but also in the field have been presented by Köhler et al. (1992) for the isopod *Oniscus asellus*. Further experiments showed *O. asellus* and the isopod *Porcellio scaber* to react to a variety of environmentally relevant substances (Cd, Pb, Zn, lindane, pentachlorophenol, PCB52, benzo[a]pyrene) with an induction of stress-70 proteins (Eckwert et al. 1994, 1997; Köhler et al. 1999b). Based on laboratory studies on the combination effect of various mixtures of heavy metals to *O. asellus*, it was shown that the elevated stress-70 level was practicable as a biomarker for heavy metal contamination in various field locations (Eckwert et al. 1997, Köhler and Eckwert 1997).

In the slug *Deroceras reticulatum*, subchronic exposure to elevated concentrations of metals (Zn, Cd, Pb) and pentachlorophenol resulted in a dose–response pattern of stress-70 induction (Köhler et al. 1994, 1996a). Diplopods (*Julus scandinavius*) have also been shown to react with elevated stress-70 levels to cadmium, zinc, and organic pollutant exposure (Zanger et al. 1994; Eckwert et al. 1994; Zanger and Köhler 1996). A change in the two-dimensional (2-D) protein pattern of the stress-70 family after cadmium treatment was described by Zanger et al. (1996). Figure 2 gives an impression of the stress-70 response to cadmium in selected soil invertebrates.

In the earthworm *Eisenia fetida*, the presence of stress-70 and its induction by thermal stress was shown by Sanders et al. (1994). Investigations on the stress-70 level in collembolans (*Orchesella bifasciata*) revealed that different populations influenced by different metal concentrations in soil have developed distinctive adaptive mechanisms in their stress response (Köhler et al. 1999a), but this could not be shown for the lycosid spider *Alopecosa cuneata* (Eckwert et al. 1994). A transgenic strain of the soil nematode *C. elegans* (Fire 1986) comprising a *Drosophila* stress-70 promotor fused to an *Escherichia coli* lacZ reporter gene was used to assess metal ion (Cd^{2+}, Zn^{2+}, Hg^{2+}, Mn^{2+}, Sn^{2+}, Ag^+), lindane, tributyltin, and polluted water toxicity in laboratory tests (Guven et al. 1994; Mutwakil et al. 1997).

Stress-60 (Chaperonin, hsp60). The protein hsp60 was originally found in the mitochondria of *Tetrahymena* (McMullin and Halberg 1987), after which it was discovered in all organisms analyzed and is closely related to the bacterial GroEL protein. Hsp60 is constitutively expressed in the nucleus. After heat

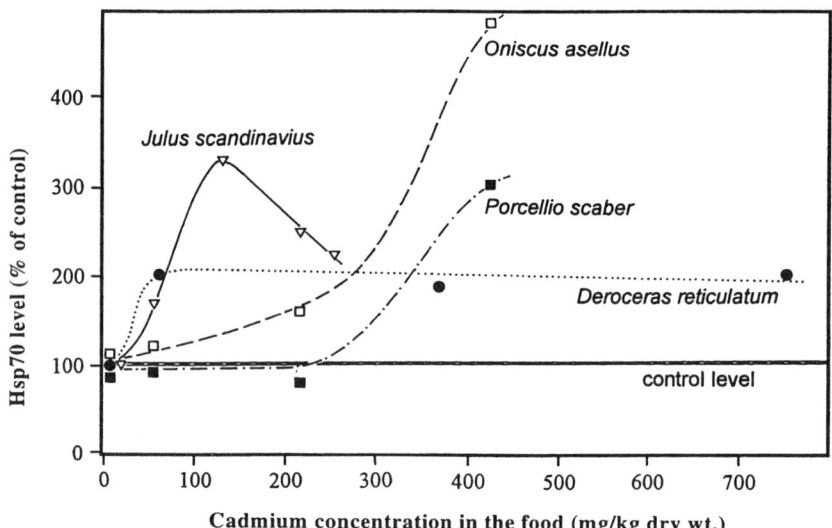

Fig. 2. Mean heat shock protein (hsp70) levels in *Deroceras reticulatum* (Pulmonata), *Oniscus asellus, Porcellio scaber* (both Isopoda), and *Julus scandinavius* (Diplopoda) in response to a 3-wk exposure to elevated cadmium (as $CdCl_2$) concentrations in their food: relative stress protein level related to controls (=100%). Note the different sensitivities of the species and the overcharge of the stress response in *J. scandinavius* at concentrations >130 mg Cd/kg food (dry weight), presumably as the result of pathological cadmium effects. Similar optimum response curves were shown also for the other species in response to other metals or metal combinations. (Data from Eckwert 1994; Eckwert et al. 1997; Köhler et al. 1996a; Zanger and Köhler 1996.)

shock, hsp60 induction may increase to about 0.3% of the total protein (Langer and Neupert 1990). It is argued that hsp60 plays an important role in the assembly of proteins in the mitochondria. The expression of the mitochondrial stress-60 protein has also been subject to biomarker studies. Sanders et al. (1994) mentioned the inducibility of chaperonin by increased temperature in the earthworm *E. fetida*. Exposure to metals (Zn, Pb, Cd), caused a slight increase in the stress-60 level in the supernatant of homogenates of *D. reticulatum, J. scandinavius, O. asellus*, and *P. scaber*, but in comparison to stress-70 the biomarker induction of metal exposure was much less sensitive (Eckwert et al. 1997). The induction of stress-60 in the nematode *Plectus acuminatus* was studied following exposure to heat, copper, and cadmium (Kammenga et al. 1998), and the hsp60 induction was related to increased concentrations of cadmium and copper. For copper, the induction was three orders of magnitude more sensitive than was the EC_{20} for reproduction. For cadmium, hsp60 induction was one order of magnitude more sensitive. The results demonstrated that hsp60 induction occurred at concentration levels that were realistic for the field situation. It was therefore suggested that hsp60 may be suitable as a potential biomarker to toxicant stress in *P. acuminatus*.

Low Molecular Weight (LMW) Stress Proteins. In contrast to stress-70 and chaperonin, LMW stress proteins are strictly stress induced but also developmentally regulated (Nover 1984; Sanders 1993). They are less conserved than stress-70 and stress-60, which results in a higher species specifity and in a minor generalization of any observed biomarker effects. In *T. pyriformis*, Amaral et al. (1988) found a set of proteins between 25 and 29 kDa to occur in response to arsenite treatment. In further laboratory studies, the promoter of the gene *hsp16*, which encodes for a 16-kDa stress protein in *C. elegans*, was fused to *E. coli lacZ* in three transgenic strains of this nematode (Stringham et al. 1992; Stringham and Candido 1993) and the animals were subjected to different stressors. The metal salts Cd, Cu, Hg, Mn, Pb, Zn, metal-polluted river water, the herbicide paraquat, and arsenite were shown to activate the hsp16 promoter in different organs of these *C. elegans* strains (Stringham and Candido 1994; Mutwakil et al. 1997; Dennis et al. 1997).

Current methods to investigate the enhanced expression of stress proteins in terrestrial invertebrates include standardized (semi)quantitative Western blot techniques, the use of transgenic animals, and the (semi)quantification of the stress protein-encoding mRNA by polymerase chain reaction (PCR). Within the field of environmental toxicology, the appropriateness of an enzyme-linked immunosorbent assay (ELISA) has been sporadically reported only in fish and centipedes (Fader et al. 1994; Vijayan et al. 1997; Pyza et al. 1997). Most of the studies cited used one-dimensional (1-D) or 2-D Western blot techniques for proteotoxicity assessment, which are actually the preferred methods in this field. The advantage of these techniques is based on the fact that they are applicable to animals taken directly from their natural habitat, which integrates such factors as possible hyperadditive effects of multiple intoxication, bioavailability of hazards, and evolved tolerance of distinct populations, all prerequisites for risk assessment in the ecosystem. Any method inaccuracy is largely surpassed by the individual variability in stress protein production, which necessitates investigation of a large number of parallel samples. The same is true for PCR studies that (semi)quantify the expression of *hsp* genes by standardized amplification of cDNA raised by reverse transcription of the hsp mRNA. Basic research on the use of this technique in toxicity assessment has been carried out for the aquatic rotifer *Brachionus plicatilis* genomic DNA (Cochrane et al. 1994) and cDNA of the slug *D. reticulatum* (Köhler et al. 1998).

In contrast, the use of transgenic animals is restricted to prospective laboratory toxicity tests and to the retrospective analysis of soil or water samples, or their extracts. Although transgenic nematode strains may be integrated in an "early warning system," it should always be borne in mind that artificial biosensors such as these do not reflect the situation of natural populations in the environment. Up to now, information on the stress protein status of populations of soil invertebrates taken from their natural habitat is scarce, but some aspects are promising. Köhler et al. (1992) found an elevated hsp70 level in an *O. asellus* population living on a smelter's spoil bank in comparison to a population from a pristine site. In a comparative study on 14 populations of the diplopod *Julus scandinavius*, a correlation of the lead and zinc concentrations in soil versus the hsp70 level in abundant *J. scandinavius* could be found for all sites except for a military area (which presumably was additionally influenced by pollutants other than metals) and two long-term polluted sites near a former opencast mine, where adaptation mechanisms likely have evolved in resident populations (Fig. 3 (summarized in Kammenga 1998). Tolerance was also demonstrated for the isopod *O. asellus* living in the aforementioned mining area; both mortality and stress-70 levels in surviving specimens in response to metal-polluted leaf litter taken from the mine site were far lower in the population living close to the mine than in a control population from a pristine site also subjected to the mine site litter (Eckwert and Köhler 1997). Observations like these may lead to the assumption that solely laboratory-based stress protein studies may be limited in their predicative potency in respect to long-term contaminated areas.

The stress protein response of terrestrial invertebrates is especially suitable to indicate the effects of exposure to comparatively low concentrations of toxicants. Like other biochemical biomarkers that require the potency of intact mechanisms to react to stressors, the stress protein level also increases first with increasing proteotoxicity, but decreases when strong adverse impacts presumably inhibit transcription or protein formation. This response could be shown for the isopod *O. asellus* (Eckwert et al. 1997) and the diplopod *J. scandinavius* (Zanger and Köhler 1996), but also for the transgenic *C. elegans* (Guven et al. 1994). To comprehend also higher yet still sublethal levels of hazardous factors, investigations on stress proteins should be combined with cellular biomarkers that constitute pathological symptoms of exposure.

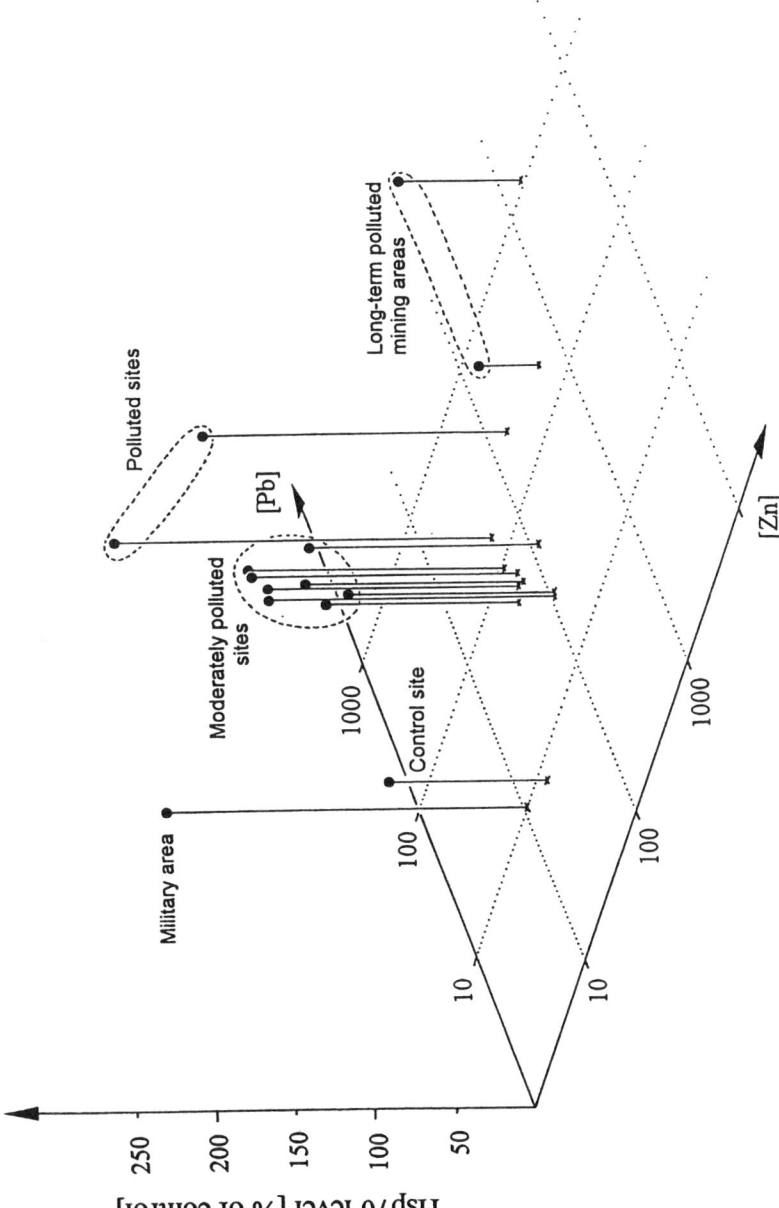

Fig. 3. Mean hsp70 levels in different *Julus scandinavius* (Diplopoda) populations taken from 14 field sites in southwestern Germany with different soil metal burdens. Stress responses were related to the concentrations of lead and zinc at the border between the soil horizons O and Ah only; however, high concentrations of cadmium were also present in the chronically polluted mining sites. [Data from M. Zanger and H.-R. Köhler; summarized in Kammenga (1998).]

III. Metallothioneins and Other Metal-Binding Proteins

Metallothioneins (MTs) are low molecular weight, cysteine-rich proteins with a high affinity to certain soft metal ions such as zinc, copper, and cadmium (Kägi and Schäffer 1988). The metals are bound to the protein moiety by cysteine sulfur atoms, being characteristically arranged in metal–mercapto clusters located in separate protein domains (Kägi 1993, Dallinger et al. 1999). Since their first discovery in the kidney cortex of the horse (Marghoshes and Vallee 1957), MTs have been detected in a variety of animal species (Hamer 1986). Due to their common structural and functional properties, MTs are now defined on the basis of a natural classification system that considers both molecular phylogenetic and structural relationships (Binz and Kägi 1999). It is widely accepted that MTs are multifunctional proteins primarily involved in homeostasis of essential trace elements, such as zinc and copper, and detoxification of excess amounts of copper, cadmium, and possibly other toxic metal ions (Cherian and Chan 1993; Dallinger et al. 1997). It has, moreover, been shown that MTs can perform a variety of additional functions, such as protection of cells against oxidative stress by radical scavenging (Sato et al. 1993) and zinc-mediated gene regulation (Zeng et al. 1991).

So far, a variety of studies has focused on structural and functional features of vertebrate MTs (Kägi 1993). In most mammalian species, for instance, low concentrations of MTs are constitutively expressed, probably in order to meet trace element demands under normal physiological conditions (Studer et al. 1997). Additionally, the synthesis of MTs can quickly be induced by events of stress that organisms suffer because of increased intracellular concentrations of trace elements, organic chemicals, or free radicals (Sato et al. 1993). In most species soft metal ions such as cadmium, zinc, or copper are among the most potent inducers of MT synthesis (Hamer 1986). As a consequence, it has long been suggested that metal detoxification would be the most important biological function of these proteins.

Several years ago, MTs were discovered in terrestrial invertebrates. In the fruit fly *Drosophila melanogaster*, for example, multiplications of MT-encoding genes have been found to be responsible for increased cadmium tolerance of certain strains of this species (Lastowski-Perry et al. 1985). MT sequences were also obtained from respective cDNAs of the nematode *C. elegans* (Imagawa et al. 1990; Slice et al. 1990). On the basis of their deduced primary structure, however, these proteins are only distantly related to vertebrate MTs.

Apart from fruitflies and nematodes, the terrestrial invertebrate MTs most thoroughly investigated are those discovered in soil-dwelling snails and slugs. Because these animals belong to the most efficient metal-accumulating terrestrial invertebrates (Dallinger 1993), it was suggested that MTs would have to play a major role in their trace element metabolism. This is true for cadmium, which is detoxified in the midgut gland of snails (*Helix pomatia, Arianta arbustorum*) and slugs (*Deroceras reticulatum*) by efficient binding to a specific MT isoform. Until now, three cadmium-binding variants of this isoform have been analyzed by their primary structure (Dallinger et al. 1993; Berger et al. 1995b). More recent findings indicate that in these organisms MTs are also involved in the homeostatic regulation of copper, possibly in connection with haemocyanin synthesis (Dallinger 1996; Dallinger et al. 1997). In this context a specific, copper-binding isoform has been isolated from the snail mantle. The primary structure of this isoform differs considerably from that of the aforementioned cadmium-binding isoforms (Berger et al. 1997).

It is well known that in vertebrates MT synthesis can be induced by metals, organic chemicals, or other events of stress. It has also been reported that MT confers protection against oxidizing agents and free radicals, being "sacrificed" in a process of thiolate oxidation and metal release (Thornalley and Vašák 1985; Thomas et al. 1986). Moreover, MT concentrations in vertebrates can be modulated by different intrinsic factors such as growth and development. Because of their similarity to vertebrate MTs, we have to assume that MT concentrations in terrestrial snails would be modulated by the same factors that interfere with MT levels in vertebrates. In contrast to vertebrates, however, snails possess organ- and metal-specific MT isoforms, which seem to react to extrinsic environmental stimuli in different manners. The midgut MT isoforms of terrestrial snails, for instance, can rapidly be synthesized in response to elevated concentrations of cadmium in the environment (Berger et al. 1995a), with induced isoforms becoming nearly exclusively loaded with cadmium. On the other hand, the concentration of the copper-binding MT isoform in the snail's mantle cannot be induced either by cadmium or by elevated concentrations of copper (Dallinger et al. 1997).

It is the strong and differential responsiveness of these different MT isoforms that make snail MTs a multiple biomarker system potentially exploitable for risk assessment in terrestrial ecotoxicology (Dallinger 1996). Because of possible species-specific peculiarities in MT structure and function it is suggested, however, to only use MTs from snail species that have thoroughly been investigated for biomarker purposes. These are, at the moment, the two pulmonate species *Helix pomatia* and *Arianta arbustorum* (Berger and Dallinger 1993). As mentioned, these snails strongly react to cadmium as an inducing agent by rapid induction of MT synthesis in their midgut gland, with a concentration increase of the cadmium-binding MT pool being linearly proportional to the cadmium concentration in this tissue (Dallinger et al. 1997). A cadmium saturation assay (Bartsch et al. 1990) has been adopted by means of which the cadmium-binding MT pool in the midgut gland of the snails can specifically be detected and quantified (Berger et al. 1995a) The accuracy and reproducibility of this method is in the same range as spectrophotometric quantification of MT on the basis of cadmium-mercapto groups (Dallinger et al., unpublished data). The MT concentration in the midgut gland of the mentioned terrestrial gastropod species can therefore be utilized as a specific measure of cadmium exposure under both laboratory and field conditions. Because (Cd)-MT can accumulate in the snail's midgut gland over extended periods of time, its concentration is a biomarker not only of recent intoxication but also for events of cadmium exposure that the animal may have experienced a long time before to the measurement (Fig. 4).

Although the synthesis of midgut gland MT isoforms can be induced by cadmium, no such induction is seen for the copper-specific mantle isoform, either by exposure of animals to elevated concentrations of cadmium, or after dosing of copper through the snail diet (Dallinger et al. 1997). However, this latter MT isoform seems to be susceptible to ionizing radiation as well as some organic chemicals that can cause a substantial reduction of MT concentration in the mantle (Fig. 4). This result is consistent with observations that under certain conditions of stress caused by free radical exposure MTs can be consumed in a process of radical scavenging (Thomas et al. 1986). Apart from this, other events of stress such as exposure to cold can also lead to reduced MT levels in the snail mantle (Dallinger et al., unpublished data). Decreasing concentrations of (Cu)-MT in this tissue may therefore indicate nonspecifically disturbing events to which animals may have been exposed by adverse climatic conditions, ionizing radiation, or any kind of xenobiotic effects causing oxidative stress. The level of (Cu)-MT in the mantle can specifically be determined by means of a modified metal saturation assay (called the tetra-thiomolybdate assay) in which copper is first replaced by cadmium and the protein is subsequently quantified as (Cd)-MT, using the known cadmium saturation approach (Klein et al. 1990).

In summary, the MT status of the snail could be expressed in terms of concentration levels of midgut gland (Cd)-MT and mantle (Cu)-MT. It remains to be assessed how variable these parameters would be under normal conditions, and to which extent they may be meaningful for biomarker purposes under varying conditions of stress and toxicant exposure in the field (see Figure 4).

Apart from terrestrial gastropods, MTs have also been detected in some earthworm species such as *Eisenia fetida* (Suzuki et al. 1980). Primary structures of earthworm MTs recently, have been derived from cDNA sequencing for the lumbricid species *Lumbricus rubellus* (Stürzenbaum et al. 1998) and *E. fetida*

FACING PAGE.

Fig. 4. *Upper boxes*: Symbolic drafts of metallothionein isoforms from midgut gland (*left hand*) and mantle tissue (*right hand*) of the Roman snail, *Helix pomatia*. The circles in the isoform chains symbolize single amino acid residues, with black circles indicating cysteine residues, white circles other amino acid residues, and grey circles in the mantle isoform representing positions for substituted residues in comparison to the midgut gland isoform. Also indicated are the most prominent properties of the two MT isoforms with respect to induction and to factors modulating MT concentration. *Middle boxes*: Quantification methods used for detection of MT concentrations, with the Cd-Chelex Assay for assessment of (Cd)-MT (left hand), and the Tetra-Thiomolybdate (TTM) Assay for quantification of (Cu)-MT (right hand). *Lower boxes*: Biomarker relevance of MT quantification in the midgut gland (left hand), and the mantle tissue (right hand), with rising concentration levels ($\uparrow$) of midgut gland MT, $[Cd-MT]_{Mi}$, and decreasing concentration levels ($\downarrow$) of mantle MT, $[Cu-MT]_{Ma}$, serving as possible stress parameters for both MT tissue pools.

Helix pomatia

Midgut Gland: Cd-MT isoform

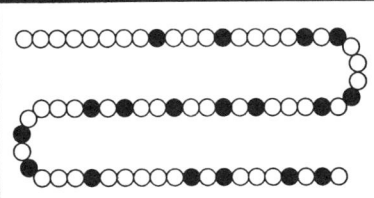

Properties:

* inducible by Cd
* MT concentration proportional to Cd exposure
* MT levels persistent over extended time period

Quantification:
Cd-Chelex Assay

Biomarker Relevance:
* Short-term biomarker for acute Cd exposure
* Long-term biomarker for chronic Cd exposure

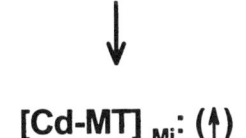

[Cd-MT]$_{Mi}$: (↑)

Mantle: Cu-MT isoform

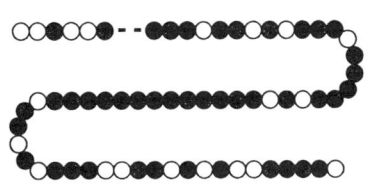

Properties:

* not inducible by Cd or Cu
* decreasing MT concentration due to ionizing radiation and oxidative stress
* decreasing MT concentration due to cold exposure

Quantification:
TTM Assay

Biomarker Relevance:
* Biomarker for ionizing radiation and oxidative stress
* Biomarker for physiological stress ?

[Cu-MT]$_{Ma}$: (↓)

(Gruber et al., unpublished data) and from amino acid sequencing for *E. fetida* (Gruber et al., unpublished data). The protein isolated from *E. fetida* possesses 12 cysteine residues, 3 of which are arranged in a triple cysteine motif (Cys-Cys-Cys). Interestingly, the cDNA sequences elucidated for *L. rubellus* and for *E. fetida* code for a putative MT that possesses more than 80 amino acids, whereas the protein isolated from *E. fetida* tissues consists of only 41 amino acids, showing a molecular weight of nearly 4200 Da. The concentration of the MT protein from *E. fetida* strongly increases upon induction by cadmium exposure, binding up to 65% of the cadmium entering the animal tissues.

The apparent discrepancy in size between the MT peptide from *E. fetida* and the putative protein encoded by the corresponding gene can only be explained by assuming that the isolated peptide must have attained its final size and shape by protease activity during the process of preparation or by deliberate posttranslational modification. Such posttranslational modifications are apparently not restricted to earthworm MTs as similar problems were recently encountered during the isolation of an MT from the insect species *Orchesella cincta* (Hensbergen et al. 1999). Upon dietary exposure to cadmium, two cadmium-binding, cysteine-rich peptides were isolated, and mass spectrometric analysis revealed that the molecular masses of these peptides were 2.9 kDa and 4.1 kDa, respectively. Amino acid sequencing of the 2.9-kDa peptide resulted in a sequence typical for a metallothionein. The results suggested that the identified peptides are products of one gene and that the primary gene product was subject to posttranslational processing. The deduced amino acid sequence of the *O. cincta* MT showed low sequence similarity with MTs from *Drosophila*. The similarity between *O. cincta* MT and MTs of invertebrates was not greater than that between *O. cincta* and vertebrates.

A cadmium-inducible, cysteine-rich protein has also been decoded from cDNA of a terrestrial enchytraeid, *Enchytraeus buchholzi* (Willuhn et al. 1994). *In vitro* translation of the transcribed product resulted in a 33-kDa protein with a high molar proportion of cysteine residues, similar to that observed in true MTs. Interestingly, the primary structure of the novel *E. fetida* MTs clearly showed close relationships to the decoded protein from *E. buchholzi*, although this latter protein is much larger in comparison to the *E. fetida* MT. Although most molecular properties of these novel proteins such as their induction potential as caused by varying extrinsic and intrinsic factors are so far unknown, their inducibility by cadmium makes them potential biomarkers for environmental pollution in the future. However, that some of these proteins are apparently subjected to posttranslational alterations limits their potiantial use as biomarkers, at least so long as the real background of these structural alterations remains unknown.

Some species of terrestrial invertebrates seem to possess, instead of MTs, non-MT, metal-binding proteins of different structure and origin. For several years attempts to prove the existence of MTs in terrestrial isopods have been without success (Donker et al. 1990). Instead, a cadmium-binding protein was isolated from metal-exposed isopods that is probably a glycoprotein involved in metal storage or detoxification (Dallinger 1993). This finding is not particularly surprising because the occurrence of metal-binding glycoproteins has also been reported from other invertebrate species (Clubb et al. 1975; Dohi et al. 1983). It remains to be clarified in the future whether such proteins might be useful as biomarkers for environmental pollution.

IV. Histology and Ultrastructure

Histological and ultrastructural changes may also serve as biomarkers and are effective tools to evaluate the toxicity of environmentally relevant chemicals in a predictive as well as in a retrospective fashion, especially when combined with biochemical markers (Triebskorn et al. 1997). Light and electron microscopy are used to diagnose cellular and subcellular symptoms of injury resulting from intoxication by xenobiotics. Both techniques are applied to locate cell death symptoms (Bowen 1981; Sparks 1972) as well as to disclose reactions to sublethal and chronic exposures in tissues and cells (summarized by Moore 1985; Storch 1988; Triebskorn 1995). However, cytotoxicity as an endpoint for environmental diagnosis is influenced by a variety of endogenous and exogenous parameters, e.g., the nutritional and the developmental state, or the sex of the test organism, the temperature, or the humidity of the environment (Braunbeck and Storch 1989; Fischer and Molnar 1992; Hryniewiecka-Szyfter and Storch 1986; Neumann 1985; Segner and Juario 1986; Segner and Braunbeck 1990; Storch 1988; Štrus et al. 1985; Triebskorn et al. 1998). Therefore, cytotoxicity refers to a cellular status quo that is the result of several factors, including the possible toxic impact of a xenobiotic not only as the single but often a quite important parameter. Moreover, in many cases, cellular reactions in animals exposed to polluted conditions differ not qualitatively but quantitatively from those occuring during the "normal" metabolic state (Köhler and Triebskorn 1998). This is not surprising, because the modification of metabolic pathways in a cell, reflected by a distinct subcellular organization, e.g., the quantity or the distribution of distinct organelles, or the ultrastructure of the organelles themselves, should be expected to be simple and less energy expensive for an organism than the reinvention of a completely new pathway. As a further problem in cellular diagnosis, in most cases cells of a distinct organ do not react simultaneously within time, and therefore "hot spots" of cellular injury appear that are not homogeneously distributed throughout the respective organ or tissue.

To overcome these aforementioned disadvantages of histological and ultrastructural biomarkers in all cases of cellular diagnosis, the "control status," i.e., the diversity of cellular reactions and the plasticity of ultrastructure occurring in an organ under natural conditions must be well known before cytotoxicological studies are started. The combination of light and electron microscope investigations is indispensable. Histological studies permit a gross check of large areas of a respective organ for pathological changes and allow one to select areas suspected of ultrastructural damage, which in turn can be analyzed in a second step with the electron microscope.

In medical research, including toxicology and human pathology, both histology and electron microscopy have been prominent as diagnostic methods. Cellular techniques, however, have entered into the field of environmental risk assessment mainly during the last two decades, whereby most of the work has focused on fish liver diagnosis (see Braunbeck and Völkl 1993; Hinton et al. 1978; Schramm et al. 1998; Sylvie et al. 1996; Triebskorn et al. 1997). In addition to fish, some marine invertebrates (mainly *Mytilus edulis, Crassostrea edulis*, and *Littorina littorea*) became standard organisms for cellular pollution monitoring (Cajaraville et al. 1990; Lowe et al. 1981; Marigómez et al. 1990, 1996; Moore 1988; Orbea et al. 1999; Köhler and Triebskorn 1998). Work on terrestrial invertebrates, however, is still scarce in this discipline (Sparks 1985; Triebskorn et al. 1991; Pawert et al. 1996), and the studies are mainly restricted to laboratory experiments with distinct xenobiotics. Furthermore, the few papers available do not focus on a particular group of organisms or on a target organ. Therefore, in the following paragraph several groups of terrestrial invertebrates are highlighted in which cellular diagnoses have been conducted using different organs or cells.

In the terrestrial gastropods *Deroceras reticulatum, Arion ater*, and *Helix pomatia*, cellular alterations have mainly been described in the hepatopancreas after molluscicide (Triebskorn 1989; Triebskorn and Künast 1990; Triebskorn et al. 1998, 1999) or metal (cadmium, mercury, or zinc) exposure (Recio et al. 1988; Marigómez et al. 1996; Triebskorn and Köhler 1996). In this organ, resorptive and basophilic cells show the most pronounced concentration–response relationships and are therefore recommended as monitor cells for environmental pollution (Cajaraville et al. 1990; Triebskorn et al. 1989; Triebskorn and Köhler 1996). In the skin and gut of slugs, however, mucocytes have been discussed as a useful cell type to describe molluscicide impact (Triebskorn and Ebert 1989; Triebskorn et al. 1998). Bourne et al. (1991) and Triebskorn et al. (1996) described ultrastructural alterations to occur in the epithelial cells of the crop after molluscicide intoxication. In these cells apical surfaces were destroyed and lipid stores drastically reduced, reflecting the enhanced energy requirement for mechanisms of detoxification (Fig. 5).

In annelids (*Eisenia fetida, Enchytraeus* spp.), the influence of several insecticides (Cypermethrin, Parathion) on cells of the chloragogenous tissue has been described by Fischer and Molnar (1992), Hagens and Westheide (1987), and Westheide et al. (1989). Additionally, Hagens and Westheide (1987) have studied the influence of parathion on gut cells in the earthworm species. An impact of the pesticide monocrotophos on oogenesis in the earthworm *Eudichogaster kinneari* was shown histologically by Lakhani et al. (1991).

In three species of mites (*Tetranychus urticae, Nothrus silvestris, Rhysotritia duplicata*), the effects of several agrochemicals (Dinoseb-Acetate, Nikkomycin, 5-fluoruracil, Apholate) or heavy metals (lead and cadmium) on gut cells (Ludwig et al. 1991; Mothes-Wagner et al. 1990), integument (Mothes 1981; Mothes and Seitz 1982), and cells of the genital tract (Langenscheidt 1973, 1975) have been shown.

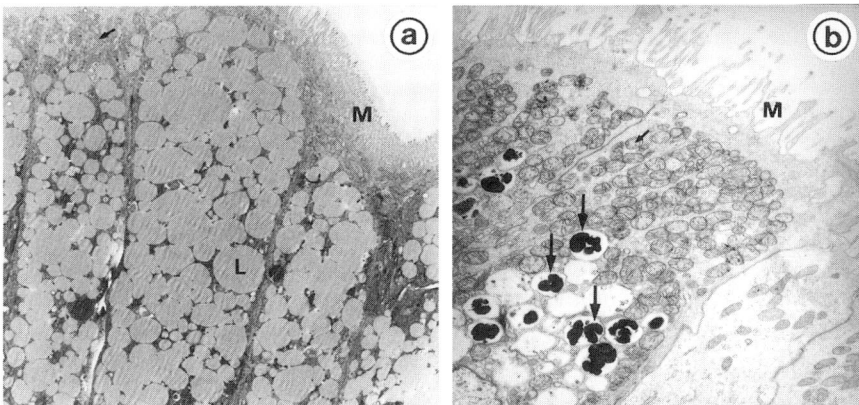

Fig. 5. (a) Overview of the crop epithelium of a control slug (*Deroceras reticulatum*). The epithelial cells are characterized by an apical microvillous border (*M*), apically located mitochondria (*arrow*), and by numerous lipid droplets (*L*) dispersed throughout the cytoplasm. ×3000. (b) Overview of the crop epithelium of a slug (*Deroceras reticulatum*) 30 hr after ingestion of 1% carbamate molluscicide cloethocarb (2-2-chloro-i-methoxy-ethoxy-)phenyl-*N*-methylcarbamate). Mitochondria (*arrowhead*) show no reaction, the amount of storage lipid is drastically reduced, and electron-dense lipofuscin particles (*arrows*) can be found distributed throughout the cell cytoplasm. Microvilli (*M*) are irregularly shaped or fused. ×8000.

As mentioned for gastropods previously, the hepatopancreas in woodlice (*Armadillidium vulgare, Oniscus asellus, Porcellio scaber*) was also found to be the most important monitoring organ to demonstrate heavy metal impacts (Prosi et al. 1983; Prosi and Dallinger 1988; Köhler et al. 1996b).

In diplopods (*Glomeris marginata, Craspedosoma alemannicum, Mycogona germanica, Polydesmus angustus, Julus scandinavius, Cylindroiulus silvarum, Tachypodoiulus niger, Ommatoiulus rutilans*), however, midgut and hepatic cells have been described to be suitable for heavy metal effect monitoring in the laboratory as well as in the field (Köhler and Alberti 1992; Berkus et al. 1994). This is also true for laboratory toxicity tests with collembolans (*Tetrodontophora bielanensis*) (Pawert et al. 1994). In several other insect groups [Orthopteroidea (*Carausius morosus*), Plannipennia (*Chrysoperla carnea*), and Coleoptera (*Epilachna varivestis*)], the effects of several insecticides (fenoxycarb, azadirachtin) have been traced in cells of the fat body (Rumpf et al. 1992; Schlüter 1984; Schlüter and Seifert 1988). Rizvi and Khan (1973) have found cells of the maphigian tubules of the orthopteran *Hieroglyphus nigrorepletus* to be sensitive to several insecticides. Furthermore, Glancey and Banks (1988) and Tedesco et al. (1986) have studied the influence of juvenile hormone analog insecticides on oocytes and ovary cells of red fire ants (Hymenoptera) and *Drosophila melanogaster* (Diptera). Finally, midgut cells of alkali bees (*Nomia mel-*

andesi) (Hymenoptera) have been described to show distinct reactions to agrochemicals (Chlorzyclizine, SKF 525A) (Moradeshagh et al. 1974).

At the subcellular level, ultrastructural investigations have revealed cell organelles to possess different susceptibilities to pollutants and to show a different spectra of reaction to them (Köhler and Triebskorn 1998). Mitochondria, for example, are generally sensitive to many kinds of stressors and react very quickly with global pathological symptoms, such as swelling, shrinkage, disruption of cristae, or formation of intramitochondrial crystals (Ghadially 1988; Triebskorn 1989; Triebskorn and Künast 1990). These alterations are often the result of general interactions of toxicants with membrane components (Reich et al. 1981) and changes in ion transport along the mitochondrial membrane (Réz 1986). Reactions of the rough and smooth endoplasmic reticulum, however, also appear very quickly after many kinds of intoxication, but compared to the reactions of the mitochondria, they are more differentiated and allow one to distinguish easily between early and progressed cell responses. Alterations such as dilation, degranulation, or vesiculation, as well as the formation of concentrically arranged whorls of the cisternae, have often been described (Ghadially 1988; Triebskorn and Köhler 1992) and have been related to functional modifications such as the induction of oxidative biotransformation enzymes (Braunbeck and Völkl 1991; Hugla et al. 1995), to the expression of recombinant integral membrane proteins (Gong et al. 1996), or to the presence of mannitol oxidase (Large and Connock 1994).

Depending on the type of stressor, the cell nucleus might react with typical symptoms either very quickly or with delay with symptoms such as darkening or brightening of the karyoplasm, dilation of the nuclear envelope, formation of specific inclusions as lipid in the karyoplasm, condensation of the chromatin, or with karyolysis (Triebskorn 1989; Vogt 1991; Vogt et al. 1994). These alterations might be caused by binding of toxicants to the chromatin, but can also be related to alterations in membranes or to any other metabolic changes leading to cell death (Vogt et al. 1994).

The Golgi apparatus generally appears to be less sensitive to intoxications, and reactions that occur are delayed compared to those of the endoplasmic reticulum. However, in the mucus cells of slugs, this organelle was very reactive and sensitive (Triebskorn and Ebert 1989; Triebskorn et al. 1998). The decrease or increase in the number of the Golgi cisternae, which may either be dilated or compressed (Somlyo et al. 1975; Triebskorn 1991; Triebskorn and Künast 1990), in addition to changes in the shape or size of the produced secretory granules (Flickinger 1971), can be interpreted as an interference of toxicants with an altered inflow of membrane material from the endoplasmic reticulum (Réz 1986). To evaluate ultrastructural stress reactions, a protocol can be used that integrates qualitative and quantitative aspects (Köhler and Triebskorn 1998). The suitability of this protocol in toxicity assessment was demonstrated for several soil invertebrates exposed to heavy metals under laboratory conditions. In these animals, species-specific differences of cellular response patterns were obtained, whereas no metal-specific patterns were found for zinc, cadmium, and lead exposure (Köhler and Triebskorn 1998).

In this review, cellular reactions to different environmental factors are considered to reflect the metabolic state of a cell, which in turn may be influenced in differentiated ways by diverse xenobiotics. Therefore, not a single symptom of a distinct organelle, but the totality of all organelle reactions that build up a syndrome of intoxication, must be used as a biomarker to interprete the influence of stress factors at the cellular level. Because of the differential sensitivities of the diverse organelles to different toxicants, this syndrome, in fact, shows some specificity for distinct substance classes.

V. Isozymes

A large number of enzymes consist of several different forms, each having different physical and physicochemical characteristics. These different forms are called isozymes and can be distinguished, for instance, by their differences in electric change and to some extent in the molecular configuration as revealed by protein gel electrophoresis (electromorphs). If the electromorphs are determined by a single gene, the enzyme consists of different allozymes (alleles of a specific gene). Several studies have shown a correlation between the nature and number of allozymes and the presence of environmental pollutants, although these studies were mainly performed with aquatic organisms (Guttman 1994). Comparatively few studies have been conducted on terrestrial invertebrates. Allozyme frequencies of glutamate-oxaloacetate transaminase in the springtail *Orchesella cincta* showed a correlation with metal tolerance (Frati et al. 1992). Another study on the springtail *Orchesella bifasciata* (Tranvik et al. 1994) did not reveal a correlation between metal concentration and the isozymes studied, glucosephosphate isomerase and phosphoglucomutase. The authors suggested that the resolution of the technique used, one-dimensional gel electrophoresis, was too low to distinguish isozyme activity in a resistant population from activity in a nonresistant population without prior experimental exposure to high metal concentrations. However, this conclusion might have depended strongly on the type of enzymes studied and the methods used. Furthermore, other factors either in the species investigated or in the ambient environment might have caused ambiguous results. Recent results showed effects of metals on one of the alleles of the enzyme glucosephosphate isomerase in the springtail *Hypogastrura persimilis* (Sørensen, personal communication) as revealed by one-dimensional electrophoresis.

Apart from heavy metals, research has focused on the effects of pesticides on isozyme activity. The effect of various pesticides on the isozyme acetylcholinesterase has been known for a long time. This esterase is important for the transmission of nerve impulses, at least in vertebrates. Eserine is known as an inhibitor of acetylcholinesterase (Augustinsson 1961) and has been used for identification of this specific esterase. Not all soil invertebrates have eserine-sensitive esterases as in *Eisenia fetida* and *Eisenia unicolor* (Øien and Stenersen

1984). Acetylcholinesterase was not found in the German cockroach, *Blatella germanica*, but the presence of cholinesterases was demonstrated (Prabhakaran and Kamble 1993). The activity of esterases was increased in strains of the cockroach resistant to chlorpyrifos, propoxur, and cypermethrin. The resistance against fenitrothion in the cotton aphid *Aphis gossypii* appeared to result from the activity of carboxylesterase (Suzuki et al. 1993). The qualitative variation in the amount of esterase and the electromorphs of esterases in the aphids were correlated with exposure to different insecticides (Furk and Hines 1993).

A response against malathion and eserine was found to be sexual, as determined in *Anopheles stephensi* (Riandry 1993). An extensive study of the esterases responsible for the organophosphate resistance in mosquitoes of the *Culex pipiens* complex has been performed during the past 20 years. Close linkage between two loci responsible for esterase production was revealed (Pasteur et al. 1981). The mechanism for this resistance was dependent on which strain of *Culex pipiens* was used for the experiments. In some cases resistance was correlated with an increase in oxidative metabolism and the enzymes associated to this pathway and no involvement of esterases (Raymond et al. 1986). Another mechanism for the organism to obtain increased resistance could be achieved either by producing a modified and more efficient enzyme or by an increased production of the enzyme—either by an increased mRNA expression rate or by multiplication of the gene responsible for the enzyme. Amplification of the gene was found for *C. pipiens* (Raymond et al. 1991). The stability of the amplification and hence the resistance was stable for at least 60 generations (Raymond et al. 1993). Detoxifying esterases were also found in *Culex quinquefasciatus* (Raymond et al. 1987); a close linkage among the loci responsible for the resistance was demonstrated (Wirth et al. 1990) and increased activity of the esterases was observed (Peiris and Hemingway 1993).

Gel electrophoresis of isozymes for visualization combined with determination of activity may be a versatile tool for risk assessment as it has been shown that various genotypes may respond differently to the presence of stressor toxicants (Benton and Guttman 1992; Rankevich et al. 1996). Several of the enzymes involved in detoxification processes display a low substrate specificity, such as esterases which can use an array of different substrates (Harris and Hopkinson 1976). The combination of gel electrophoresis and enzyme activity is shown in Figure 6, which illustrates the sensitivity of the esterase isozymes from *Eisenia fetida* when exposed to dimethoate, a cholinesterase inhibitor. These results might lead one to postulate that changed activity, changed specificity, or an altered expression of enzymes may be among the first responses to a stressor. For the highly specific enzymes such as glucosephosphate isomerase and phosphoglucomutase, however, both enzymes necessary for the glycolytic pathway, the response to environmental stress was not unambiguous (Tranvik et al. 1994).

The potential number of different invertebrate isozymes for risk assesment of soil pollutants may be vast, as it is now possible to detect more than a hundred isozymes (Manchenko 1994). When analytical procedures and laboratory

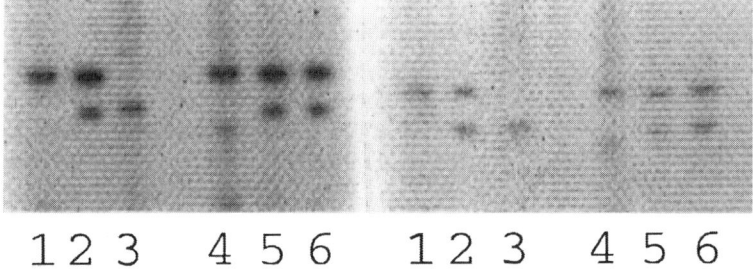

Fig. 6. Esterases from six individuals of *Eisenia fetida* (1–6) stained for esterase with α-naphthylactate and Fast Blue RR salt for 30 min at 37 °C for isoelectrifocusing. *Left*: Esterase zymogram from the six individuals obtained by using the normal staining procedure. *Right*: Reduced activity of the esterase isozymes from the same six individuals when 0.01 M dimethoate is added to the staining solution.

facilities are established and optimized a great number of individuals can be analyzed per day at a reasonable cost, but a necessary assumption is that an unambiguous relationship between the nature or activity of the isozyme and the toxicant exists. At present the application of isozymes for soil risk assessment is still in its infancy and requires more attention, because the potential for using it as a method for monitoring is evident. It is envisaged that, in general, isozymes can be used either as a specific biomarker of exposure to cholinesterase inhibitors (esterases) or as general indicators for toxic stress because individuals exposed to toxicants will compensate for any damage by using possibilities for a changed metabolism before using other compensatory methods (e.g., reduction in reproduction, reduction of lifetime).

VI. Lysosomal Membrane Integrity

At the subcellular level, the lysosomal system has been identified as a particular target for the toxic effects of xenobiotics (Moore 1990). Lysosomes are a morphologically heterogeneous group of membrane-bound subcellular organelles containing acid hydrolase that range in diameter from 250 Å to 1 µm. The function of lysosomes in the cells is to catabolize organelles and macromolecules. A change in lysosomal membrane stability is thought to be a general measure of stress. In stable lysosomes, hydrolases are prevented from reacting with substrates by an intact membrane. Membrane stability decreases in response to stress as membrane permeability increases. The mechanism causing this alteration is not well understood and may vary with the type of stressor. Pathological alterations in lysosomes have been especially useful in the identification of adverse effects for a range of organisms, with much evidence for aquatic organisms (Moore 1990; Svendsen and Weeks 1995) but rather limited evidence for terrestrial species (Weeks and Svendsen 1996).

In earthworms it was possible to measure the effects of both inorganic and organic contaminants by determining changes in the permeability of the lysosomal membrane of earthworm coelomocyte cells using the neutral red retention assay as described by Weeks and Svendsen (1996). Neutral red is an established technique for the evaluation of cellular toxicity. It has been used to test the cytotoxicity of chemical compounds, pharmaceutical agents, surfactants, food additives, pesticides, solvents, complex toxicant mixtures, and a variety of miscellaneous chemical agents using cells from mammals, humans, and fish as targets. The neutral red retention time (NRRT) assay makes use of the fact that healthy lysosomes take up and retain the dye (neutral red) indefinitely, whereas dye in lysosomes of coelomocyte cells taken from "stressed" earthworms gradually leaks into the surrounding cytoplasm. This response has been quantified in both the laboratory and field using various organic and inorganic contaminants (Svendsen et al. 1996; Harreus et al. 1997; Svendsen and Weeks 1997a,b). This marker is nonspecific, responding equally sensitively to organic or inorganic contamination; however, if used in combination with an earthworm immunocompetence marker (Svendsen et al. 1998), such as total immunoactivity of the coelomocytes, then it may be possible to be more specific as to the likely nature of contamination.

Changes in lysosomal integrity as a result of chemical stress have been investigated for a range of earthworm species including the lumbricid species (*Lumbricus terrestris, L. castaneus, L. rubellus,* and *Aporrectodea rosea*) and in *Eisenia andrei*. At present few chemicals have been tested, although the list is growing. To date, most studies have examined the effects of copper (Svendsen and Weeks 1997a,b; Harreus et al. 1997) and only a few have examined the effects of organic contaminants (Svendsen et al. 1996; Svendsen et al. 1998). For example, clear dose–response relationships were seen in earthworms collected from soil contaminated with pyrolyzed plastics after an industrial accident (Svendsen et al. 1996). Eason et al. (1999) also observed effects for organic compounds such as polyaromatic hydrocarbons (PAHs).

Exposure to copper-contaminated soil in both the laboratory and field mesocosm studies showed a clear dose–response relationship for the lysosomal membrane stability with *L. rubellus, A. rosea,* and *E. andrei* (Svendsen and Weeks 1997a,b). Exposure to increasing soil copper concentrations resulted in a concomitant decrease in NRRT (Fig. 7). A similar dose–response relationship was obtained for *E. veneta* exposed to different concentrations of sewage sludge (Olesen 1998). Although few studies have been performed, correlations between NRRT and reproductive output and mortality have been observed for *L. rubellus* and *E. andrei*, both in laboratory and in field mesocosm studies (Svendsen and Weeks 1997a,b).

Lysozyme activity in earthworm coelomic fluid has been found to be a potential biomarker for assaying the immunotoxicity of metals (Goven et al. 1994). Many experimental studies with marine invertebrates have shown that these lysosomal alterations can be induced by single toxicants such as copper and PAHs (Moore et al. 1988). However, similar patterns of generalized lysosomal

A
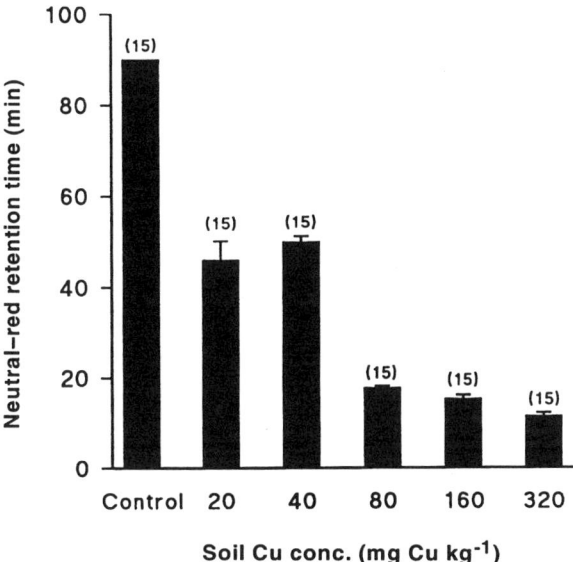

B
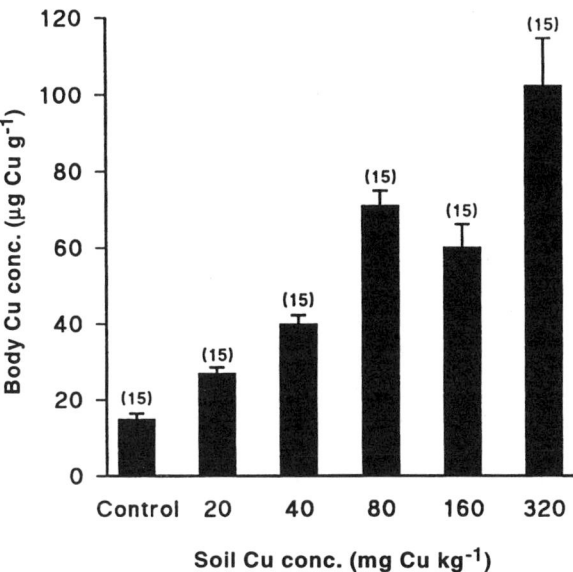

Fig. 7. (A) Mean (±SEM) neutral red retention times (min) and (B) body copper concentrations (μg g^{-1} dry wt.) measured in earthworms (*Eisenia andrei*) exposed to an increasing range of soil copper concentrations for 28 d under laboratory conditions. (*n*) are given in parentheses (after Svendsen and Weeks 1997a).

response have been observed to be induced by nonchemical stressors such as hypoxia, hyperthermia, osmotic shock, and dietary depletion (Moore 1985). Thus, it would appear that many adverse conditions are capable of inducing such changes. This nonspecificity of the lysosomal reactions is therefore of great value as a general indicator of deterioration in the health of the animal and hence its environment.

VII. Novel Biomarkers

Ecotoxicological biomarkers have been studied predominantly for vertebrate species, but with increasing emphasis on invertebrates and in particular aquatic invertebrates. Despite the number of investigations carried out on these biomarkers, few direct relationships between activity and exposure have been elucidated. Numerous examples of biomarkers can be found in the literature (even for soil invertebrates) in addition to the biomarkers reviewed here. Because many biomarkers are general indicators for (toxic) stress, there is a need for further investigation to study novel approaches to biomarker elucidation and development. In this section we focus on the potential of using nuclear magnetic resonance (NMR) spectroscopy for discovering new biomarkers.

The primary role of NMR spectroscopy has been in elucidation of the physical structure of organic chemicals, although advances in instrumentation and computing capabilities, increased sensitivity, resolution, chemical shift dispersion, and dynamic range have allowed broadening in the scope of its applications (Nicholson and Wilson 1989). In addition there has been a trend toward hyphenated NMR techniques, with systems such as HPLC-NMR-MS (mass spectroscopy) being extremely powerful in these complex studies. An area of particular interest has been in biological NMR spectroscopy, where biochemical processes have been revealed and measured in many types of complex biological matrices including biological fluids, tissue extracts cells, and living tissues. In particular, high-resolution proton NMR spectroscopy has been applied to investigate the composition of vertebrate biofluids in a variety of studies on endogenous and xenobiotic metabolism (Nicholson and Wilson 1989).

Proton NMR spectroscopy has many features making it particularly useful in metabolic and toxicological research. It has thus proved to be successful as a biomedical probe as it provides a rapid, specific but nonselective (i.e., exploratory) and nondestructive detector for a wide range of low molecular weight compounds in small samples of biological fluid or tissue extracts. Toxicological studies involving NMR spectroscopy have led to the detection of novel biomarkers of toxicity and disease in mammals, especially with the aid of pattern recognition methods (Nicholson et al. 1985; Nicholson and Wilson 1989). In mammalian toxicology it is more important to understand the subtle chronic effects rather than the acute toxic effects, which may also apply to studies with invertebrates. Unfortunately, NMR spectroscopy does have some disadvantages such as its relative insensitivity in comparison to some other spectroscopic techniques, with the detection limits being in the region of 5 µM for proton NMR spectroscopy at 500 MHz using a standard 5-mm NMR probe. In addition, the high initial purchase price of a spectrometer is often prohibitively expensive, up to $2.5 million US, which has tended to restrict their usage to studies on vertebrates. In comparison to studies on vertebrates and in the medical field, only a handful of studies on NMR spectroscopy on invertebrates are found in the literature.

NMR spectroscopic techniques that have been applied to soil invertebrate biochemical systems have been concerned with the structure elucidation of a small range of steroids or proteins in highly purified samples (Lee and Aarhus 1993). Gibb et al. (1997a) used high-resolution proton NMR methods to characterize the low molecular weight components in tissue extracts of whole invertebrate representatives found in terrestrial environments and with known pollution indicator potential. This approach was used with a view to developing NMR pattern-recognition (PR) techniques for investigating metabolic profile perturbations in ecotoxicological studies. The authors looked at tissue extracts from a range of earthworm, isopod, and mollusc species and showed that each extract contained highly complex mixtures of low molecular weight metabolites in control samples. Despite the high degree of signal overlap observed, many resonances were assigned. Obvious differences in the spectra for the different species were observed not only in the ratios of certain metabolites but also in the presence or absence of certain metabolites. Furthermore close phylogenetic similarities for related species were found.

Traditionally, the use of pollution indicator species has involved first, the measurement of accumulated pollutant residue and second the determination of lethal dose concentrations. Subsequently, critical dose levels are assessed to investigate the more realistic sublethal effects of pollutants. It is, however, at the cellular and subcellular levels that xenobiotics have their effects, and a more diagnostic approach would be to investigate biochemical changes resulting from exposure to pollutants. Homeostasis maintains metabolites within certain ratios; with toxic insult, however, there is often loss of this control and a subsequent increase or decrease in particular metabolites. Characterization of such metabolite "biomarkers" has been classically achieved by specific biochemical assaying. This approach tends to be time consuming and labor intensive. In contrast, NMR requires little sample pretreatment and is rapid and not preselective. Gibb et al. (1997a) suggested that NMR fingerprints closely reflect the metabolic status of the individual because soil invertebrates have unique and characteristic fingerprints of endogenous metabolites in their basal metabolic state.

As a wide range of biochemicals in important intermediary pathways can be measured simultaneously by proton NMR spectroscopy, the potential for the detection of biochemical perturbations is high. Gibb et al. (1997b) applied high-resolution proton NMR spectroscopy to investigate the biochemical effects of Cu(II) in two earthworm species, *Eisenia andrei* and *Lumbricus rubellus*, exposed under different laboratory and field conditions. The most marked metabolic response was the elevation of endogenous whole-body free histidine in animals, which correlated positively with increasing copper exposure and total copper burden in earthworms from a semifield study (Fig. 8). Histidine forms thermodynamically very stable copper complexes under a wide range of conditions. The authors proposed that the elevation of free histidine in response to copper challenge provided energetically "low-cost" copper detoxification mechanisms. It was suggested that the use of low molecular weight chelators in metal detoxification strategies may be more widespread than previously thought. Histidine may thus act as a biomarker of such exposure.

It may be concluded that there is scope for the development and understanding of novel biomarkers of exposure and effect from the use of techniques such as proton NMR spectroscopy.

VIII. Confounding Factors and Transiency of the Biomarker Response

The most important limitations, inherent in the use of biomarkers as indicators for toxic stress, are (i) the sensitivity of the biomarker response to various natural biotic and abiotic conditions, which are referred to as confounding factors, and (ii) the transiency of the biomarker response. After exposure to contaminants, the initial induction of some biomarkers may fade away, thus obscuring the applicability of biomarkers in the field to assess the history of exposure or effect. On the other hand, other biomarkers may be induced for extended periods as long as the exposure continues, thus enhancing field applicability. Because studies on the confounding factors and the transiency of biomarker responses in soil invertebrates are scarce (Lagadic et al. 1994), this section also touches upon these aspects in aquatic organisms.

The response of hsps and MTs to toxic compounds may be influenced by natural fluctuations in environmental conditions such as temperature, soil moisture content, and availability of food. If an organism is collected from the field, and its natural environment is extreme for one or more conditions other than contamination, as a consequence the biomarker may be poor to analyze environmental stress because of contamination. Fader et al. (1994) pointed out that there is a need to determine the role of hsps in the physiological and ecological response of organisms to normal environmental changes such as seasonal variation in temperature and photoperiod. This information seems to be important because it is well known that hsps are induced by a wide variety of physical and chemical agents such as metals (cadmium, zinc, mercury, lead), pesticides (i.e., carbonates, organophosphates, and paraquat), dioxins, oxidizing agents, UV irradiation, temperature, and food shortage (Sanders 1993). The complexity and variability as found in the field therefore also tends to confound the establishment of causal links between hsp induction and contamination of biological systems. This problem is exemplified in a study by Pedersen et al. (1997) on the hsp60 and hsp70 response to organotin concentrations in mussels (*Mytilus edulis*) collected from different sites around the island of Fyn, Denmark. The hsp70 levels did not vary significantly between sites, whereas the hsp60 levels

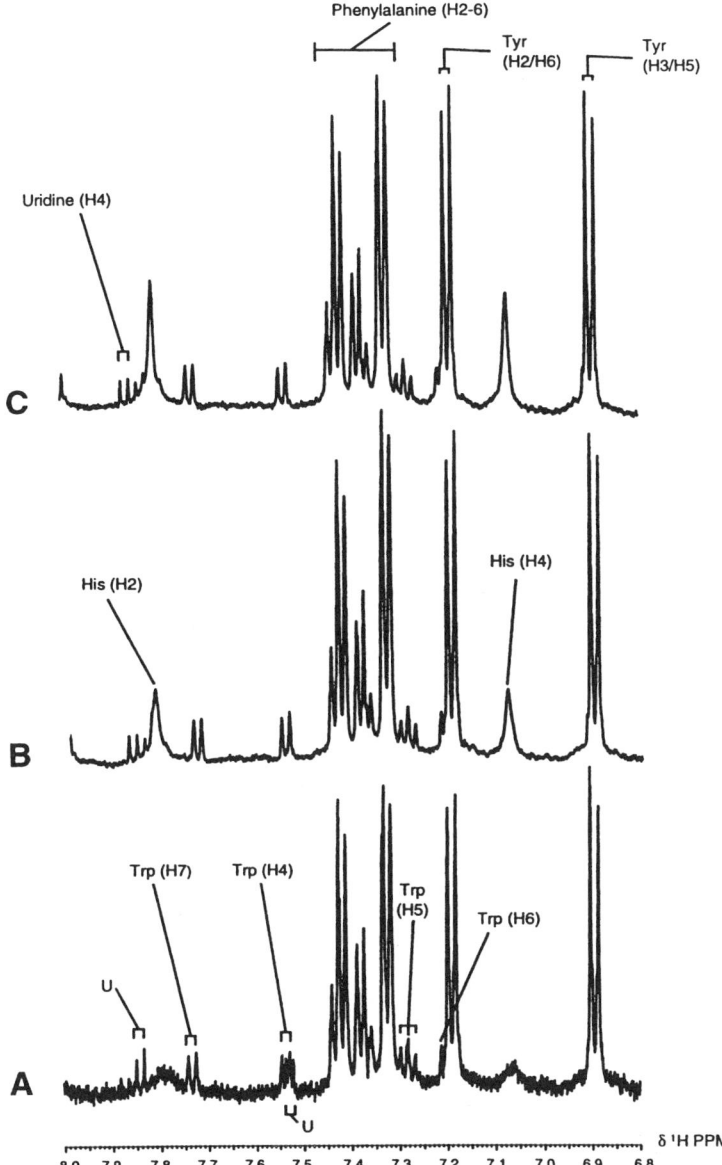

Fig. 8. 500-MHz ^{1}H-NMR spectra (partial region only) of earthworm (*Lumbricus rubellus*) tissue extracts from control (A) and after exposure to 160 mg Cu kg^{-1} soil dry wt (B). The latter sample was spiked with authentic histidine (C) to give a final concentration of approximately 1.24 mM. *His*, histidine; *Trp*, tryptophan; *Tyr*, tyrosine; *U*, unassigned. Bracketed abbreviations denotes position of protein [number of scans (NS) = 256, line broadening (LB) = 1]. (After Gibb et al. 1997b.)

were significantly elevated in mussels from one of the sites. However, the stress protein levels did not reflect the degree of organotin contamination of mussel tissues, indicating that interpretation of stress protein biomarker responses requires further detailed elucidation on confounding factors before they can be usefully incorporated in routine environmental monitoring programs. An example of the transciency of the hsp response was reported in a study on bluegill sunfish exposed to contaminated sediment. Western blots of hsp70 from the gills indicated that the stress protein level was present in higher concentrations in exposed fish after 2 wk. However, there were no differences in amount of hsp70 from week 14 to week 16, and the amount of hsp70 in exposed fish was even less than in controls (Theodorakis et al. 1992). The fact that hsps were synthesized within the period of a general response to ambient and internal stress suggests that hsp induction was probably transient, and thus an exclusive application of hsps as a biomarker should be carried out with caution. Vedel and Depledge (1995) and Köhler et al. (1999b) suggest that hsp70 may have limited value because of the transient nature, particularly after exposure to chemicals with a log P_{ow} of about 3.5 or higher. Given these problems concerned with confounding factors and transciency, it is recommended not to apply the hsp biomarker solely but rather to include it in a suite of other biomarkers. The inclusion of hsps in a biomarker set is necessary if the aim is also to register the adverse effects of nonchemical stressors (e.g., temperature), which in concert with the chemical pollution determine the actual and therefore ecologically relevant stress situation in field sites.

Like hsps, MTs may also be induced by various other stress factors or are regulated by hormonal control at the cellular level (Benson and DiGiulio 1992). The induction of MT appears to be strongly influenced by the ambient temperature, diet, and reproductive conditions (Stegeman et al. 1992), and unless a specific tissue can be identified for the purpose, variability in protein levels associated with naturally occurring events such as molting and growth or reproductive activity are likely to mask toxicant-induced MT responses. This limitation was illustrated by George and Olsson (1994), who reviewed the applicability of MT as a sublethal biomarker of exposure to cadmium, copper, lead, and zinc. It appeared that the hepatic MT concentration of the fish *Limanda limanda*, sampled along a gradient from the Elbe estuaries to the Dogger bank in the North Sea, did not reflect the sedimentary metal concentrations along this gradient. However, branchial MT concentrations and the concentrations of copper, cadmium, and zinc displayed a clear relationship with sedimentary metal concentrations. The hepatic MT levels of the females were influenced by sexual maturation (by a factor of 7), and the male levels mainly depended on internal copper and zinc status but did not reflect sedimentary metal levels.

Another aspect that may obscure a proper interpretation of the MT response is the intrinsic biological control of MT metabolism. For instance, the rate of biodegradation of MT is dependent on the amount of metal bound; MT induced in the rat liver by copper, zinc, or cadmium had a half-life of 12–16, 20, and

70–90 hr, respectively (Feldman and Cousins 1976; Bremner et al. 1978). Moreover, both copper and zinc were lost when MT biodegraded; cadmium, however, remained bound to MT in the steady state during both biosynthesis and biodegradation (Ridlington et al. 1981). Similar observations have been made for the blue crab, in which cadmium concentrations in the digestive gland and the gill remained constant for a period of 12 d after cadmium exposure was stopped (Brouwer et al. 1984). Also, molting frequency may interfere with MT response in some insects and crustaceans. Relationships between levels of MT and and the ecdysteroid molting hormone have been observed during the molt cycle of the blue crab *Callinectus sapidus* (Engel and Brouwer 1987). According to Benson and DiGiulio (1992), the use of MT as a biomarker may be premature so long as the role of confounding factors and its tranciency are not elucidated.

The MT expression in the springtail *O. cincta* is temperature dependent in a known manner (Donker et al., unpublished data). Springtails are organisms that reach an equilibrium internal metal concentration after about three molting cycles. After reaching their equilibrium, the MT levels at 20 °C are about half of the expression as found at 10 °C. The transiency of the response is high. *O. cincta* was exposed to cadmium for more then 12 weeks, and from the moment *O. cincta* reached its equilibrium cadmium concentration, the MT levels in the springtail were independent of the exposure time (Donker et al., unpublished). This reliable and transient response makes this animal very suitable for biomarker studies.

Much information is available with regard to the influence of confounding factors and the response transiency of isozymes. Decreases in temperature, e.g., in winter time, changed the activity of certain enzymes and distribution between electromorphs of the isozyme aldolase in the gall moth *Epiblema scudderiana* (Joanisse and Storey 1995). However, increasing temperatures during spring showed that the process was reversible. The change in the activity of enzymes may lead to an altered metabolism, which ultimately may reduce the effect of a change in temperature on an invertebrate. Another example on transciency was the change in the water balance, which was found for the locust *Locusta migratoria*, when exposed to various insecticides. This effect disappeared a few hours after the insecticide injection (Proux et al. 1993).

Esterase inhibition is caused by phosphorylation of the enzyme, which is not rapidly reversible. In vertebrates, the blood concentration of esterase appears to be far from constant as the result of temporal, diurnal, or developmental variation (Thompson et al. 1988; Walker 1995). In addition, changes in temperature and feeding regimes have marked effects on enzyme activity in fish (Jimenez et al. 1988) and insects (McDonald et al. 1990). This problem may be overcome by measuring specific esterases, or even different isozymes, using a method based on purification of specific esterases with antibodies and quantification by ELISA or by using densitometry combined with electrophoresis (Benton and Guttman 1997). Temperature was shown to have an effect on allozymes of glucosephosphate isomerase in the isopod *Porcellio laevis*; the allele with the slower migration rate was favored in "cool" conditions, but the allele that migrated faster was favored in stressed environments (McCluskey et al. 1993). Higher genetic diversity was found for the snail *Pomatias olivieri* living under stressed conditions caused by temperature and hence increased risk for desiccation (Rankevich et al. 1996).

Lysosomes are known to be linked to pathological changes in animals, and these changes are associated with a variety of diseases induced by environmental pollutants. Also, lysosomal responses have been observed to be induced by nonchemical stressors such as hypoxia, hyperthermia, osmotic shock, and dietary depletion (Moore 1985). The NRRT assay measures the efflux of Neutral Red out of the lysosomes into the cytosol; the retention time being an integrated reponse to membrane damage. Studies on the role of confounding factors are still in their infancy. As yet, fish and earthworm lysosomes appear to be relative insensitive to confounding factors (Mayer et al. 1992; Svendsen et al. 1996).

Another aspect that may hamper a proper interpretation of the biomarker response is the adaptability of organisms to various pollutants. On an evolutionary time scale, transiency of esterase activity may vary significantly. Esterases functioned as a detoxification mechanism in *Culex pipiens* when the insects were exposed to organophosphorous compounds, and long-term exposure resulted in an amplification of a gene determining esterase (Raymond et al. 1991). The stability of the amplification was very high even when insecticide usage was reduced (Raymond et al. 1993). Crossing strains of *C. pipiens* that contained the amplification with strains without amplification showed that amplification of the gene was only stable in homozygous strains.

Under conditions of chronic exposure, as is the case for metal pollution in soil, the metal toxicity may be accomodated by behavioral or physiological acclimation with no apparent effect on growth, reproduction, or survival. Alternatively, the stress may differentially affect various genotypes within the population, resulting in progressive elimination of sensitive individuals and a shift in the genetic structure of the population. It has been shown that chemical stressors exert strong evolutionary pressures on plant and animals. Survival of a fraction of the population following repeated or extreme exposure to chemical stress may lead to genetic selection of a part of the population capable of stress avoidance, stress detoxification, or repair or compensation for injury, thereby increasing their capacity for survival (Ernst and Peterson 1994). The adaptive capacity of the organism may determine the response of the biomarker, which will depend on the degree of adaptation to environmental stress and on the homogeneity of the investigated population. Populations are composed of a great diversity of genotypes, each having different potential to react to contaminants in the environment. As a result, the fittest genotypes will survive and bring the population to a adaptation level differing from that before the environmental change. In a situation where the organisms are permanently exposed to the toxicants, a population may exist of largely adapted genotypes. The biomarker response of these individuals may differ from that of the "original" population.

IX. Ecological Relevance of Biomarkers in Soil Invertebrates

The response of soil invertebrate biomarkers to toxic stress only has ecological relevance if the response can be linked to either (i) exposure levels of the toxicant (biomarker of exposure) or (ii) effects (biomarker of effect) at higher organization levels such as the population or community level. Biomarkers of exposure reflect the bioavailability of the toxic compounds in soils, and a suite of these kinds of biomarkers provides insight into the exposure to the "total toxic load." A suite of biomarkers consisting of general markers for stress (hsps, lysosomal integrity) and specific markers for metals (metal-binding proteins) or some pesticides (esterases) may well mirror the exposure to different toxicants at the same time. An ecologically relevant biomarker of effect may be an early warning for any lethal or sublethal effect at the individual level or could be directly related to the functioning of the organism at higher organization levels. Most research on biomarkers focuses on understanding the responses at lower levels of biological organization such as molecules, cells, and individuals. However, contaminants exert their effects at all levels of biological organization from molecules to ecosystems. It is unlikely that a substantive linkage can be made between lower-level biomarker responses and changes in populations, communities, or ecosystems. Not surprisingly, very little research has been conducted to demonstrate any causal relationship between such biomarker responses and damage at higher orders of biological organization.

Effect on populations are mediated through effects on individuals and therefore biomarker responses should be linked to parameters at the individual level. To link the response of any biomarker to a particular change at the individual level, one should ideally know the whole cascade of molecular and biochemical pathways leading to the observed effect. The biomarker response may be part of this pathway. Such an approach may prove to be very difficult, however, because the biochemical routes often are unknown. It is therefore advocated that one seeks for probable correlations between a biomarker reponse and an individual or population-level effect. An example of such an approach was given by Köhler et al. (1998). They studied the induction of stress gene transcription (including mRNA stability) and the accumulation of the corresponding stress protein hsp70 in slugs exposed to cadmium- or zinc-enriched food. To validate the suitability of these two aspects of the cellular stress response to act as early-warning markers for metal effects on life history parameters, the fecundity, offspring number, longevity, and mortality of slugs were recorded in life cycle experiments. Köhler et al. found that the elevation of the hsp70 mRNA level caused by short-term (14-d) metal exposure coincided with both diminished fecundity and reduced offspring production as the result of chronic metal exposure in terms of threshold concentrations for cadmium and zinc effects. Similar work was carried out by Migula (1999) on the regression of phosphofructokinase activity in the pea aphid *Acyrthosiphon pisum* reared on alfalfa exposed to SO_2 for 2 wk on female development rate. These two examples indicate that there can be interesting relationships between biomarker responses and demographic parameters under laboratory conditions. To go beyond the population level in the field seems a little too far fetched at the moment, because one should recognize the biomarker response transiency, its normal fluctuation under ambient conditions, and the influence of climatic variations and life stage. A promising study in this direction was reported by Svendsen et al. (1996) on the use of lysosomal memebrane integrity in natural earthworm populations exposed to industrial pollution in a forest floor ecosystem.

In conclusion, it is required to understand the induction and nature of a biomarker response, its significance in terms of its immediate environment, and the confounding factors before attempting to link its predicitve value to population-level effects. It is both easier and cost-effective to measure a biomarker at the lower end of any organizational cascade. The establishment of a dose–response relationship is imperative; it is also far easier to interpret such a response when it is measured as indicated in Fig. 1. If we can establish a significant effect at these low levels, we should be confident that such an effect can be predicted with time to occur at all levels of any continuum—the important variable being the temporal difference. This understanding circumvents the need to measure ecosystem damage directly, a task that is both costly and time consuming. More appropriate is the need to have reliable predictive diagnostic biomarkers as tools to link biomarker responses at the lower level to effects on any higher level (Weeks 1998).

X. Biomonitoring Using Soil Invertebrate Biomarkers

In this general section we attempt to provide a stepwise approach for the implementation of a biomarker monitoring survey using soil invertebrate species as put forward by Weeks et al. (1999).

A. Chemical Residue Measurement

The determination of chemical residues in selected compartments (soil, pore water, biota) is a fundamental starting point in any overall survey procedure. Data from such measurements inevitably pinpoint the requirement to make a biomarker selection and further refined chemical residue measurements. If there is no detectable residue, this may not necessarily indicate the absence of toxicity exerted by a mixture of different compounds. Biomarkers for general stress may provide more insight into a combined toxicity effect. In addition, there may still be a requirement to undertake biomarker analysis as there may well be secondary transformation and metabolism of the parent compounds.

B. Choice of Taxa and Complementary Activities

The choice of taxa largely depends on the availability of species present with sufficient abundance in the area where the study or monitoring scheme is to be carried out. The alternative is the use of introduced indicator species or species transplanted from clean environments to monitor the biomarker response, particularly in areas where suitable organisms do not occur naturally or have been eradicated through severe pollution. Where ever possible, noninvasive samples should be taken.

Furthermore, temporal analysis of soil animal community assemblages may be conducted noting the effect of seasonal changes. This initial survey will show obvious disparities in species composition for former versus actual faunal analyses, thus guiding the final selection of appropriate biomarker/bioindicator species but also foretelling changes to ecosystem function.

It is important in the first instance to determine aspects of soil health such as soil processing and functionality, including turnover of nitrogen and carbon, decomposition rates, and soil pH. When assessing the effects of chemicals in terrestrial systems, not only should measurements at a structural level be taken (e.g., changes in population and community composition), but also information at the soil process level is required. Soil processes such as organic matter decomposition are determined by the activity of the resident fauna mediating this processes and soil microbial activity, are key measurements in any survey, because they truly reflect the "status" of the soil. We recommend the following measurements be undertaken:

Litter decomposition: Using the litter bag or litter container techniques with the most relevant plant species in the area; it is advisable to compare exposed and nonexposed vegetable material of two different types of plant species widespread in the area. Because decomposition of organic matter is the main gateway of essential elements to the soil, information given by these measurements is important to understand the processes of nutrient cycles in this area. Connected to the measurements of mass loss over time, measurements of chemical residues, nutrients, and also some biological parameters (e.g., fungal biomass and litter respiration) are advisable, allowing a more rounded comprehension of the processes occurring.

Soil respiration: Respiration is a key ecosystem function, giving information about total soil activity. This can be measured by SIR (substrate-induced respiration) or basal respiration, comparing polluted and reference representative soils. Soil microbial biomass can also be estimated by this method. Preliminary assays to assess the natural fluctuations in these parameters should be done. Continuous measurements taking into account the temperature regime should be made whenever possible.

Microbial and faunal activity in soil: These parameters can be measured using the well-known "cotton-strip" and "bait lamina" methods (Kratz 1998). The first is a good indicator of microbial activity, and the latter has produced promising results in measuring feeding activity of soil fauna, especially in forested areas. Measurements must be made over time, particularly when the soil fauna is more active (seasonality).

Reconnaissance: Before a full-scale biomarker-based monitoring program is started, some form of reconnaissance or exploratory study should be carried out to provide information on the maximum distance at which the pollutant(s) can be detected above ambient levels, so that sampling sites can be positioned to cover the extent of pollution deposition. This preliminary survey should also attempt to collect information as to the distribution of the chosen biomarker monitoring organisms(s) and hence the distribution of sampling sites. It would also provide information on the practicalities that are essential for planning a full-scale monitoring program, such as the number of subsamples required to give the desired level of precision.

Sampling strategy: Where, when and how many? As mentioned, the sampling sites should be chosen to represent the whole area over which the pollutant in question can be said to have originated from the source being monitored. In addition, sampling areas immediately outside the area where the pollutant can be detected is useful for warning of cumulative impacts or progression of continued pollutant loading. The number of sampling points deployed depends on the resources available. Obviously however, the more points sampled the greater the accuracy of the biomarker measurements, the better likelihood of detecting change, and the greater reliability (statistical accuracy) of the data. Subsampling at each station will increase the precision of pollutant estimation. At the simplest level, a transect of 10 to 12 points along a downwind gradient of the source (point source) should be sufficient to show the distance over which the pollutants are likely to travel, be deposited, and exert biomarker responses, and thus allow accurate curve fitting. Possibly the optimum sampling scenario would be to monitor sites at various and increasing distance from the source in all directions; these could be stratified to include different community types. The effects of seasonality of sampling must be known and the representativity of samples collected understood.

Sampling methods: The most commonly employed methods should be adopted for the species or group of animal selected during the initial survey and reconnaissance. All reasonable precautions should be taken to prevent further contamination or disturbance of the collected samples. Samples should be taken from similar situations (habitat). Samples should be taken away from local sources of potential contamination that may give unrepresentative readings. A representative subset of samples should be taken at each site to give a measure of precision. It would be more informative to measure both a biochemical/subcellular/cellular marker in association with a physiological/reproductive biomarker for the species selected.

Statistical analysis and detection of change: For simple transects it may be possible to compare best-fit lines for different sets of data. For more comprehensive surveys it may be enough to compare best-fit lines for a number of transects separately. Problems will arise if different transects show different trends. Where measures of community composition have been made, then the use of multivariate measures, especially canonical correspondence analysis, could be used to identify the influence of different pollutants and environmental factors.

XI. Future Potential and Limitations for Soil Risk Assessment Using Biomarkers

The following are the current soil and water ecotoxicological risk assessment methods for chemical compounds as adopted by the European Union (Ahlers and Diderich 1998):

- Exposure assessment: leading to predicted environmental concentrations (PEC) of a chemical from release by its production, processing, use, and disposal.
- Effect assessment: data obtained from acute or long-term toxicity tests are used to extrapolate concentrations with expectedly no adverse effects on organisms (no effect concentration, NEC).
- Risk characterization: the PEC is compared with the NEC. When the PEC > NEC, an attempt should be made to revise data of exposure and/or effect in an iterative process to conduct a refined risk characterization. When PEC remains larger than NEC, risk reduction measures must be considered.

Risk assessment tools based on soil invertebrate biomarkers have not yet been developed, and this field is still open to debate. The success of these tools for risk assessment depends largely on the choice of biomarker assays (for both effect and exposure assessment), which should both be sensitive and reflect the relevant ecological effects of toxicity at the site. The field of biomarker research has developed rapidly in the last decade and triggered an ongoing debate on the pros and cons of various biomarkers to be included in risk assessment procedures. Of great consideration is the validity of a predictive role for biomarkers in the assessment of risks to soil organism communities and the potential for the circumvention of lasting damage to their populations. In conjunction with Section VIII on confounding factors and response tranciency, this section aims to provide some provocation for establishing comprehensive links between biomarkers and soil risk assessment.

A. Biomarker Potential

Recent advances in soil invertebrate toxicity tests (Løkke and Van Gestel 1998) and biomarker tests described in this review allow one to estimate, with greater confidence, the ecological risk to certain components of the ecosystem. For example, the increasing use of relevant toxicity endpoints (e.g., reproduction) in earthworms and the use of exposure–effect biomarkers in earthworms and field exposure approaches using mesocosms have all recently emerged as appropriate technologies for the ecological risk assessment of contaminated soils in conjunction with chemical residue analyses. Many of these techniques have yet to be fully linked to each other to fulfill the requirements of a robust risk assessment framework or to monitor the change in risk during and following the remediation of polluted areas. For this purpose, biomarkers may be used in hazard identification, in exposure assessment, and to associate a response with the probability of population decline. Identification of practicable biomarkers associated with different toxicity test endpoints requires interdisciplinary cooperation and research. The use of biomarkers linked with toxicity test results should enhance this process and increase the reliability of prediction of risk. Improved prediction of the risk will permit effective intervention to protect soil ecosystems in general and polluted sites in particular.

As reviewed here, the increasing development of biomarker applications in terrestrial invertebrates for measuring exposure, effective concentrations, responses, and susceptibilities has been brought forward by the application of new analytical techniques, mainly based on the tools of molecular biology and biochemistry. The question remains, however, to what extent biomarkers can be applied to estimate or predict ecological harm at higher levels of organization, rather than as a simple biochemical measure of response. The step of measuring a selective response in a single representative organism and extrapolating from this simple biochemical change so as to understand the ecological significance of a slight increase or decrease of a measured ecosystem parameter is a major hurdle, yet one that will provide a potentially powerful legislative tool (Weeks et al. 1999).

B. Biomarker Limitations

The limitation of biomarkers in view of soil risk assessment can be found mainly in their ambigious relationship with demographic or other higher level endpoints, as pointed out in Section IX. Furthermore, the nonspecificity of certain biomarkers makes it difficult to pinpoint any toxic compound in particular, whereas confounding factors may well obscure a proper and clear interpretation of the biomarker response in terms of ecosystem health. We therefore propose that a standardized protocol for monitoring purposes must be developed for the application of a suite of biomarkers for field situations. One way forward in this direction might be to divide the results emanating from a biomarker suite into classes of severity of response to produce an index of bioavailability and hence toxicity (Triebskorn et al. 1997; Pawert et al. 1998; Schramm et al. 1998). This step will be combined with existing and evolving approaches to risk assessment and risk management to allow the user community to take bioavailability and the biological significance of toxicity into account. The decision-making protocol must draw upon ranking of biomarker responses on an arbitrarily defined scale from 0 to 5, for example, in which the predicted "impact index" 0 represents control conditions or no observable biomarker response, and the "impact index" 5 refers to conditions inducing maximum biomarker responses (Köhler and Triebskorn 1998). The ranking of scales obviously must be adjusted for every investigated marker individually. Based on this approach, a standardized protocol combining the predictive value of a suite of biomarkers will provide quantitative information on the toxic potential connected to specific exposure conditions. It is possible that such a protocol may be repeated after reclamation or remediation.

XII. Conclusions

Ecotoxicological research on biomarkers in soil invertebrates is still in its infancy, although significant progress has been made concerning the potential and limitations of various biomarkers for a large number of soil species. Table 1 summarizes the most important aspects of the various biomarkers reviewed in this article, and the following conclusions can be drawn:

The hsp response in soil invertebrates is especially suitable to indicate the effects of exposure to comparatively low concentrations for a range of toxicants and can be regarded as a biomarker of general stress. The hsp level increases first with increasing proteotoxicity, but decreases when adverse impact presumably inhibits transcription or protein formation. Reliable methods have been developed that can be applied to a range of different species and which allow for highly reproducible and trustworthy results based on hsp60 and hsp70 induction.

The application of MTs and other metal-binding proteins as biomarkers for exposure in soil invertebrates has a long-standing record, and increased information is becoming available on new MTs in more soil invertebrate species. In addition, new and rapid methods are being developed for analyzing MT induction both at the protein and the molecular level, and reliable and reproducible methods are currently available. (Cd)-MT is well characterized for the springtail (*Orchesella cincta*), the Cd concentration dependence has been proven in the laboratory. As springtails are organisms that reach an equilibrium metal concentration, its MT concentration is a useful biomarker for exposure as well as for effect, the more so because the transciency of the response and the temperature dependence have been investigated recently. The disadvantage of the measurement of *O. cincta* MT at the protein level (i.e., 100 individuals are needed per measurement) can be overcome by using quantitative RT-PCR. Because (Cd)-MT can accumulate in snail midgut glands over extended periods of time, its concentration is a biomarker not only for recent intoxication but also for events of cadmium exposure that the snails may have experienced a long time before the measurement.

Cellular and histological alterations reflect the "health" state of a cell, which may be a measure for the presence of toxicants. Histopathological work on terrestrial invertebrates, however, is still scarce. In terrestrial gastropods and isopods, cellular alterations have been described mainly in the hepatopancreas. In this organ, resorptive cells (B cells) and basophilic cells (S cells) show the most detailed responses to sublethal contaminations and are therefore recommended as monitor cells for environmental pollution. For diplopods, the midgut and hepatic cells, and for Collembola, the midgut epithelial cells, are suitable.

Isozymes have been poorly studied in soil invertebrates despite their promising role as potential biomarkers in aquatic organisms. Among the large diversity of isozymes, the best studied are esterases frequently used as biomarkers of exposure to various classes of pesticides. Many other isozymes offer poten-

Table 1. Overview of soil invertebrate biomarkers, studied under laboratory or field conditions, and their characteristics, specificities to toxic compounds and sample tissue. MT = metallothionein, MB = non-MT metal binding protein.

Type of biomarker	organism		Biomarker of:	Known specificity for a class of compounds	Sample tissue
	laboratory studies	field studies			
Hsp60 and 70	protozoans nematodes isopods gastropods diplopods earthworms	isopods gastropods springtails spiders	exposure, sensitivity	specificity is low	whole body
MT	nematodes gastropods earthworms springtail enchytraeids	gastropods	exposure, sensitivity	cadmium, zinc, copper but also to other factors	midgut gland and mantle in gastropods
MB	isopods	isopods	exposure	cadmium	
Histological and ultastructural markers	gastropods earthworms enchytraeids mites, woodlice diplopods some insects	diplopods	exposure, sensitivity	specificity depends on type of organelle	hepatopancreas, skin, gut or crop in gastropods and woodlice; chloragogenous tissue in annelids; gut in mites; midgut and hepatic cells in diplopods
Isozymes	springtails earthworms aphids mosquitoes	springtails	exposure, sensitivity	pesticides but also other compounds	whole body
Lysosomal integrity	earthworms	earthworms	exposure, sensitivity	non-specific	fluid sample
Histidine	earthworms		exposure	specific to copper	whole body

tials for biomarker research, such as glucosephosphate isomerase and phosphoglucomutase, both enzymes necessary for the glycolytic pathway. Gel electrophoresis of isozymes combined with determination of enzyme activity has been suggested as a versatile tool for risk assessment. It is envisaged that, in general, isozymes can be used either as a specific biomarker of exposure to cholinesterase inhibitors (esterases) or as general indicators for toxic stress.

The lysosomal system has been identified as a particular target for the toxic effects of xenobiotics although it has a limited application in soil invertebrates. In earthworms, it was possible to measure the effects of both inorganic and organic contaminants by determining changes in the permeability of the lysosomal membrane of earthworm coelomocyte cells using the neutral red retention assay. This marker is nonspecific, responding equally sensitively to organic or inorganic contamination; however, if used in combination with an earthworm immnunocompetence such as total immunoactivity of the coelomocytes, then it is possible to be more specific as to the likely nature of contamination.

The use of proton NMR techniques has revealed a potential new biomarker. Free histidine was positively correlated with increasing copper exposure and total copper burden in earthworms from a semifield study. Histidine forms copper complexes that are thermodynamically very stable under a wide range of conditions, and it was proposed that the elevation of free histidine in response to copper challenge provided an energetically "low cost" copper detoxification mechanisms. Histidine may thus act as a biomarker of exposure.

The transient responses and confounding factors of biomarkers can obscure a proper interpretation of biomarker responses under field conditions. These factors are still very poorly understood and require more study. For interpretation of biomarker measurements, one needs to know to what extent the biomarker expression depends on these factors. It is advocated that research in this area focus on the transiency of responses both in the laboratory and in (semi) field bioassays. Continuously exposed animals should be allowed to remain in noncontaminated conditions for specific periods in the laboratory and then analyzed for transient induction.

For risk assessment purposes, it is recommended that the aforementioned biomarkers may show promise when included in a suite of biomarkers among different soil invertebrate species. It is recommended that a risk assessment protocol must draw upon ranking of biomarker responses on a defined scale in which the predicted "lowest impact index" represents control conditions (no observable biomarker response) and the "highest impact index" refers to conditions inducing maximum biomarker responses. The ranking of scales must be adjusted for every investigated marker individually. Based on this approach, a standardized protocol combining the predictive value of a suite of biomarkers can provide quantitative information on the toxic potential connected to specific exposure conditions.

Summary

This review has served to present the most recent information on a selected series of biomarker studies undertaken on soil invertebrates during two extensive European-funded scientific consortia, BIOPRINT and BIOPRINT-II. The goals were to develop and validate methods for the analysis of markers of stress in a range of soil-dwelling organisms. We have discussed the potential and limitations of the following invertebrate biomarkers for soil risk assessment purposes: heat shock proteins, histological and ultrastructural markers, metallothioneins and metal-binding proteins, esterases, lysosomal integrity, and the novel biomarker histidine. The hsp response in soil invertebrates is especially suitable to indicate the effects of exposure to comparatively low concentrations for a range of toxicants and can be regarded as a biomarker of general stress. The application of MTs and other metal-binding proteins as biomarkers for exposure in soil invertebrates has been well described, and new methods are being developed for analyzing MT induction both at the protein and molecular level, and reliable and reproducible methods are now available. (Cd)-MT is well characterized for the springtails and its MT concentration is a useful biomarker for exposure as well as for effect. For snails, (Cd)-MT can accumulate in the midgut gland over extended periods of time and therefore its concentration is a biomarker not only for recent intoxication but also for events of cadmium exposure that snails may have experienced a long time before the measurement took place. Cellular and histological alterations can be regarded as reflecting the "health" state of a cell, which may be a measure for the presence of toxicants. Histopathological work on terrestrial invertebrates, however, is still scarce.

Isozymes have been poorly studied in soil invertebrates despite their promising role as potential biomarkers in aquatic organisms. Among the large diversity of isozymes, the most well studied are esterases that are frequently used a biomarkers of exposure to various classes of pesticides. Many other isozymes offer potentials for biomarker research, such as glucosephosphate isomerase and phosphoglucomutase, both enzymes necessary for the glycolytic pathway. The lysosomal system has been identified as a particular target for the toxic effects of xenobiotics, although it has yet a limited application in soil invertebrates. This marker is nonspecific, responding equally sensitively to organic or inorganic contamination; however, if used in combination with an earthworm immnunocompetence assay such as total immunoactivity of the coelomocytes, then it is possible to be more specific as to the likely nature of contamination. Free histidine was positively correlated with increasing copper exposure and total copper burden in earthworms from a semifield study. Histidine may thus act as a biomarker of exposure. The transient responses and confounding factors of biomarkers obscure a proper interpretation of biomarker responses under field conditions. These factors are still very poorly understood and require more study.

For risk assessment purposes it is recommended that the aforementioned biomarkers may show promise when included in a suite of biomarkers among different soil invertebrate species. It is recommended that a risk assessment protocol draw upon ranking of biomarker responses on a defined scale. It is also hoped that the problems outlined in this review will aid the direction of future research on soil invertebrate biomarkers.

Acknowledgments

The present review was written within the framework of the EU-funded research projects of the Environment (BIOPRINT, contractnr. EV5V-CT94-0406) and Environment and Climate (BIOPRINT-II, contractnr. ENV4-CT96-0222) programmes of the 4th framework.

References

Ahlers J, Diderich R (1998) Legislative perspective in ecological risk assessment. In: Schüürmann G., Markert B. (eds) Ecotoxicology. Wiley, New York, pp 841–868.
Amaral MD, Galego L, Rodrigues-Pousada (1988) Stress response of *Tetrahymena pyriformis* to arsenite and heat-shock: difference and similarities. Eur J Biochem 171: 145–160.
Augustinsson KB (1961) Multiple forms of esterase. Ann NY Acad Sci 94:844–860.
Bartsch R, Klein D, Summer KH (1990) The Cd-Chelex assay: a new sensitive method to determine metallothionein containing zinc and cadmium. Arch Toxicol 64:177–180.
Benson WH, DiGiulio RT (1992) Biomarkers in hazard assessments of contaminated sediments. In: Burton GAJ (ed) Biomarkers in Hazard Assessments of Contaminated Sediments. Lewis, Boca Raton, FL.
Benton MJ, Guttman SI (1992) Allozyme genotype and differential resistance to mercury pollution in the caddisfly, *Nectopsyche albida*. II. Multilocus genotypes. Can J Fish Aquat Sci 49:147–149.
Benton MJ, Guttman SI (1997) Electrophoretic evidence of esterase inhibition in larval caddisflies exposed to inorganic mercury. Water Environ Res 69:240–243.
Berger B, Dallinger R (1993) Terrestrial snails as quantitative indicators of environmental metal pollution. Environ Monit Assess 25:65–84.
Berger B, Dallinger R, Thomaser A (1995a) Quantification of metallothionein as a biomarker for cadmium exposure in terrestrial gastropods. Environ Toxicol Chem 14: 781–791.
Berger B, Hunziker PE, Hauer CR, Birchler N, Dallinger R (1995b) Mass spectrometry and amino acid sequence of two cadmium-binding metallothionein isoforms from the terrestrial gastropod *Arianta arbustorum*. Biochem J 311:951–957.
Berger B, Dallinger R, Gehrig P, Hunziker PE (1997) Primary structure of a copper-binding metallothionein from mantle tissue of the terrestrial gastropod *Helix pomatia* L. Biochem J 328:219–224.
Berkus M, Gräff S, Alberti G, Köhler H-R (1994) The impact of the heavy metals, lead, zinc, and cadmium, on the ultrastructure of midgut cells of *Julus scandinavius* (Diplopoda). Verh Dtsch Zool Ges 87:321.
Binz P-A, Kägi JHR (1997) Metallothionein: Molecular evolution and classification. In: Klaassen CD (ed) Metallothionein IV. Birkhäuser, Basel, pp 7–13.
Bourne NB, Jones GW, Bowen ID (1991) Endocytosis in the crop of the slug *Deroceras reticulatum* (Müller) and the effects of the ingested molluscicides, metaldehyde and methiocarb. J Molluscan Stud 57:71–80.

Bowen ID (1981) Techniques for demonstrating cell death. In: Bowen ID, Lockshin RA (eds) Cell death in biology and pathology. Chapman & Hall, London, pp 379–444.

Braunbeck T, Storch V (1989) Zelle und Umwelt: Wie wirken sich Umweltgifte auf Zellen aus? BIUZ 19:127–132.

Braunbeck T, Völkl A (1991) Induction of biotransformation in the liver of eel *Anguilla anguilla* L. by sublethal exposure to dinitro-*o*-cresol: an ultrastructural and biochemical study. Ecotoxicol Environ Saf 21:109–127.

Braunbeck T, Völkl A (1993) Toxicant-induced cytological alterations in fish liver as biomarkers of environmental pollution? A case study on hepatocellular effects of dinitro-*o*-cresol in golden ide (*Leuciscus idus melanotus*). In: Braunbeck T, Hanke H, Segner H (eds) Fish Ecotoxicology and Ecophysiology. VCH, Weinheim, pp 55–80

Bremner I, Hoekstra WG, Davies NT, Young BW (1978) Metabolism of 35S-labelled copper-, zinc-and cadmium-thionein in the rat. Chem Biol Interact 23:355–367.

Brouwer M, Brouwer-Hoexum TM, Engel DW (1984) Cadmium accumulation by the blue crab, *Callinectes sapidus*: involvement of hemocyanin and characterization of cadmium-binding proteins. Mar Environ Res 14:71–88.

Cajaraville MP, Diez G, Marigomez JA, Angulo E (1990) Responses of basophilic cells of the digestive gland of mussels to petroleum hydrocarbon exposure. Dis Aquat Org 9:221–228.

Cherian MG, Chan HM (1993) Biological functions of metallothionein—a review. In: Suzuki KT, Imura N, Kimura M (eds) Metallothionein III. Birkhäuser, Basel, pp 87–109.

Clubb RW, Lords J, Gaufin A (1975) Isolation and characterisation of a glycoprotein from stonefly, *Pteronarcys californica*, which binds cadmium. J Insect Physiol 21: 53–60.

Cochrane BJ, Mattley YD, Snell TW (1994). Polymerase chain reaction as a tool for developing stress protein probes. Environ Toxicol Chem 13:1221–1229.

Craig EA, Gross CA (1991) Is hsp70 the cellular thermometer? Trends Biol Sci 16: 135–140.

Craig EA, Weissman JS, Horwich AL (1994) Heat shock proteins and molecular chaperones: mediators of protein conformation and turnover in the cell. Cell 78:365–372.

Dallinger R (1993) Strategies of metal detoxification in terrestrial invertebrates. In: Dallinger R, Rainbow PS (eds) Ecotoxicology of Metals in Invertebrates. Lewis, Boca Raton, FL, pp 245–289.

Dallinger R (1996) Metallothionein research in terrestrial invertebrates: synopsis and perspectives. Comp Biochem Physiol C 113(2):125–133.

Dallinger R, Berger B (1993) Function of metallothioneins in terrestrial gastropods. Sci Total Environ (suppl pt 1):607–615.

Dallinger R, Berger B, Hunziker PE, Birchler N, Hauer CR, Kägi JHR (1993) Purification and primary structure of snail metallothionein: similarity of the N-terminal sequence with histones H4 and H2A. Eur J Biochem 216:739–746.

Dallinger R, Berger B, Hunziker PE, Kägi JHR (1997) Metallothionein in snail Cd and Cu metabolism. Nature (Lond) 388:237–238.

Dallinger R, Berger B, Hunziker P, Kägi JHR (1999) Structure and function of metallothionein isoforms in terrestrial snails. In: Klaassen CD (ed) Metallothionein IV. Birkhäuser, Basel, pp 173–178.

Dennis JL, Mutwakil MHAZ, Lowe KC, de Pomerai DI (1997) Effects of metal ions in combination with a non-ionic surfactant on stress responses in a transgenic nematode. Aquat Toxicol 40:37–50.

Depledge MH, Fossi MC (1994) The role of biomarkers in environmental assessment. 2. Invertebrates. Ecotoxicology 3:161–172.
De Pomerai DI (1996) Heat-shock proteins as biomarkers of pollution. Hum Exp Toxicol 15:279–285.
Dohi Y, Ohba K, Yoneyama Y (1983) Purification and molecular properties of two cadmium binding glycoproteins from the hepatopancreas of a whelk, *Buccinum tenuissimum*. Biochim Biophys Acta 745:50–60.
Donker MH, Koevoets P, Verkleij JAC, van Straalen NM (1990) Metal binding compounds in the hepatopancreas and haemolymph of *Porcellio scaber* (Isopoda) from contaminated and reference areas. Comp Biochem Physiol C 97:119–126.
Eason C, Svendsen C, O'Halloran K, Weeks JM (1999) An assessment of the lysosomal neutral red retention test and immune function assay in earthworms (*Eisenia andrei*) following exposure to chlorpyrifos and benzo-[a]-pyrene and contaminated soil. Pedobiologia in press
Eckwert H (1994) Induktion von heat shock Proteinen durch Schwermetalle: Proteinbiochemische und cytologische Grundlagen sowie angewandte Aspekte. Diploma Thesis, University of Heidelberg, Germany.
Eckwert H, Köhler H-R (1997) The indicative value of the hsp70 stress response as a marker for metal effects in *Oniscus asellus* (Isopoda) field populations: variability between populations from metal-polluted and uncontaminated sites. Appl Soil Ecol 6:275–282.
Eckwert H, Alberti G, Köhler H-R (1997) The induction of stress proteins (hsp) in *Oniscus asellus* (Isopoda) as a molecular marker of multiple heavy metal exposure. I. Principles and toxicological assessment. Ecotoxicology 6:249–262.
Eckwert H, Zanger M, Reiss S, Musolff H, Albert G, Köhler H-R (1994) The effect of heavy metals on the expression of hsp70 in soil invertebrates. Verh Dtsch Zool Ges 87:325.
Edington BV, Wheelan SA, Hightower LE (1989) Inhibition of heat shock (stress) protein induction by deuterium oxide and glycerol: additional support for the abnormal protein hypothesis of induction. J Cell Physiol 139:219–228.
Engel DW, Brouwer M (1987) Trace metal binding proteins in marine molluscs and crustaceans. Biol Bull 172:239–251.
Ernst WHO, Peterson PJ (1994) The role of biomarkers in environmental assessment: 4. Terrestrial plants. Ecotoxicology 3:180–192.
Fader SC, Yu Z, Spotila JR (1994) Seasonal variation in heat shock proteins (hsp70) in stream fish under natural conditions. J Therm Biol 19:335–341.
Feldman SL, Cousins RJ (1976) Degradation of hepatic zinc-thionein after parenteral zinc administration. Biochem J 160:583–588.
Fire A (1986) Integrative transformation of *Caenorhabditis elegans*. EMBO J 5:2673–2680.
Fischer E, Molnar L (1992) Environmental aspects of the chloragogenous tissue of earthworms. Soil Biol Biochem 24:1723–1727.
Flickinger CJ (1971) Alterations in the Golgi apparatus of amoebae in the presence of an inhibitor of protein synthesis. Exp Cell Res 68:381–387.
Frati F, Fanciulli PP, Posthuma L (1992) Allozyme variation in reference and metal-exposed natural populations of *Orchesella cincta* (Insecta: Collembola). Biochem Syst Ecol. 20:297–310.
Furk C, Hines CM (1993) Aspects of insecticide resistance in the melon and cotton aphid, *Aphis gossypii* (Hemiptera: Aphididae). Ann Appl Biol 123:9–17.

George S, Olsson PE (1994) Metallothioneins as indicators of trace metal pollution In: Kramer KJM (ed) Biomonitoring of Coastal Waters and Estuaries. CRC Press, Boca Raton, FL, pp 151–171.
Gething M-J, Sambrook J (1992) Protein folding in the cell. Nature (Lond) 355:33–45.
Ghadially FN (1988) Ultrastructural Pathology of the Cell and Matrix' 3rd Ed. Butterworths, London, pp 1–587.
Gibb JOT, Holmes E, Nicholson JK, Weeks JM (1997a) Proton NMR studies on tissue extracts of invertebrate species with pollution indicator potential. Comp Biochem Physiol 118B:587–598.
Gibb JOT, Svendsen C, Weeks JM, Nicholson JK (1997b) ^{1}H NMR spectroscopic investigations of tissue metabolite biomarker responses to Cu(II) exposure in terrestrial invertebrates: identification of free histidine as a novel biomarker of exposure to copper in earthworms. Biomarkers 2:295–302.
Glancey BM, Banks WA (1988) Effect of the insect growth regulator Fenoxycarb on the ovaries of queens of the red imported fire ant. Ann Entomol Soc Am 81:642–648.
Gong F-C, Giddings TH, Meehl JB, Staehelin LA, Galbraith DW (1996) Z-membranes. Artificial organelles for overexpressing recombinant integral membrane proteins. Proc Natl Acad Sci USA 93:2219–2223.
Goven AJ, Chen SC, Fitzpatrick LC, Venables BJ (1994) Lysozyme activity in earthworm (*Lumbricus terrestris*) coelomic fluid and coelomocytes: enzyme assay for immunotoxicity of xenobiotics. Environ Toxicol Chem 13:607–613.
Guttman SI (1994) Population genetic structure and ecotoxicology. Environ Health Perspect 102:97–100.
Guven K, Duce JA, De Pomerai DI (1994) Evaluation of a stress-inducible transgenic nematode strain for rapid aquatic toxicity testing. Aquat Toxicol 29:119–137.
Hagens M, Westheide W (1987) Subletale Schädigungen bei *Enchytraeus minutus* (Oligochaeta, Annelida) durch das Insektizid Parathion: Veränderungen in der Ultrastruktur von Chloragog- und Darmzellen in Abhängigkeit von der Belastungsdauer. Verh Ges Kol 14:237–241.
Hamer DH (1986) Metallothionein. Annu Rev Biochem 55:913–951.
Harreus D, Koehler H-R, Weeks JM (1997) Combined non-invasive cell isolation and neutral-red retention assay for measuring the effects of copper on the lumbricid *Aporrectodea rosea* (Savigny). Bull Environ Contam Toxicol 59:44–49.
Harris H, Hopkinson DA (1976) Handbook of Enzyme Electrophoresis in Human Genetics. North-Holland, Amsterdam.
Hensbergen PJ, Donker MH, Van Velzen MJM, Roelofs D, Van der Schors RC, Hunziker PE, Van Straalen NM (1999) Primary structure of a cadmium-induced metallothionein from the insect *Orchesella cincta* (Collembola). Eur J Biochem 259:197–203.
Hinton DE, Klaunig JE, Lipsky MM (1978) PCB-induced alterations in teleost liver: a model for environmental disease in fish. Mar Fish Rev 40:47–50.
Hryniewiecka-Szyfter Z, Storch V (1986) The influence of starvation and different diets on the hindgut of Isopoda (*Mesidotea entomon*, *Oniscus asellus*, *Porcellio scaber*). Protoplasma 134:53–59.
Hugla JL, Phillippart JC, Kremers P, Goffinet G, Thome JP (1995) PCB contamination on the common barbel (*Barbus barbus*) (Piscies, Cyprinidae) in the river Meuse in relation to hepatic monooxygenase activity and ultrastructural liver change. Neth J Aquat Ecol 29:135–145.

Imagawa M, Onozawa T, Okumura K, Osada S, Nishimura T, Kondo M (1990) Characterization of metallothionein cDNAs induced by cadmium in the nematode *Caenorhabditis elegans*. Biochem J 268:237–240.

Jimenez BD, Burtis LS, Ezell GH, Egan BZ, Lee NE, Mc Carthy JF, Beauchamp JJ (1988) Effects of environmental conditions on the mixed function oxidases system in blue gill sunfish (*Lepomis macrochirus*). Environ Toxicol Chem 7:623–634.

Joanisse DR, Storey KB (1995) Temeprature acclimation and seasonal responses by enzymes in cold-gall insects. Arch Insect Biochem Physiol 28:339–349.

Kägi JHR (1993) Evolution, structure and chemical activity of class I metallothioneins: an overview. In: Suzuki KT, Imura N, Kimura M (eds) Metallothionein III. Birkhäuser, Basel, pp 29–55.

Kägi JHR, Schäffer A (1988) Biochemistry of metallothionein. Biochemistry 27:8509–8515.

Kammenga JE (1995) Progress Report 1994 of BIOPRINT. Second technical report. Ministry of Environment and Energy, National Environmental Research Institute, Silkeborg, Denmark.

Kammenga JE (1996) Progress Report 1995 of BIOPRINT. Third technical report. Ministry of Environment and Energy, National Environmental Research Institute, Silkeborg, Denmark.

Kammenga JE (1998) Progress Report 1997 of BIOPRINT-II. Second technical report. Ministry of Environment and Energy, National Environmental Research Institute, Silkeborg, Denmark.

Kammenga JE, Arts MSJ, Oude-Breuil WJM (1998) HSP60 as a potential biomarker of toxic stress in the nematode *Plectus acuminatus*. Arch Environ Contam Toxicol 34: 253–258.

Kammenga JE, Simonsen V (1997) Manual of BIOPRINT-II. First technical report. Ministry of Environment and Energy, National Environmental Research Institute, Silkeborg, Denmark.

Klein D, Bartsch R, Summer KH (1990) Quantification of Cu-containing metallothionein by a Cd-saturation method. Anal Biochem 189:35–39.

Köhler H-R, Alberti G (1992) The effect of heavy metal stress on the intestine of diplopods. In: Meyer E, Thaler K, Schedl W (eds) Advances in Myriapodology. Ber Natwiss-Med Ver Innsbr (suppl) 10:257–267.

Köhler H-R, Belitz B, Eckwert H, Adam R, Rahman B, Trontelj P (1998) Validation of *hsp70* stress gene expression as a marker of metal effects in *Deroceras reticulatum* (Pulmonata): correlation with demographic parameters. Environ Toxicol Chem 17: 2246–2253.

Köhler H-R, Eckwert H (1997) The induction of stress proteins (hsp) in *Oniscus asellus* (Isopoda) as a molecular marker of multiple heavy metal exposure. II. Joint toxicity and transfer to field situations. Ecotoxicology 6:263–274.

Köhler H-R, Eckwert H, Triebskorn R, Bengtsson G (1999a) Interaction between tolerance and 70 kD stress protein (hsp70) induction in collembolan populations exposed to long term metal pollution. Appl Soil Ecol 11:43–52.

Köhler H-R, Hüttenrauch K, Berkus M, Gräff S, Alberti G (1996b). Cellular hepatopancreatic reactions in *Porcellio scaber* (Isopoda) as biomarkers for the evaluation of heavy metal toxicity in soils. Appl Soil Ecol 3:1–15.

Köhler H-R, Knödler C, Zanger M (1999b) Divergent kinetics of hsp70 induction in *Onsicus assellus* (Isopoda) in response to four environmentally relevant organic chemicals (B[a]P, PCB52, (-HCH, PCP): suitability and limits of a biomarker. Arch Environ Contam Toxicol 36:179–185.

Köhler H-R, Rahman B, Gräff S, Berkus M, Triebskorn R (1996a) Expression of the stress-70 protein family (hsp 70) due to heavy metal contamination in the slug, *Deroceras reticulatum*: an approach to monitor sublethal stress conditions. Chemosphere 33:1327–1340.

Köhler H-R, Rahman B, Rahmann H (1994) Assessment of stress situations in the grey garden slug, *Deroceras reticulatum*, caused by heavy metal intoxication: semi-quantification of the 70 kD stress protein (hsp70). Verh Dtsch Zool Des 87:328.

Köhler H-R, Triebskorn R (1998) Assessment of the cytotoxic impact of heavy metals on soil invertebrates using a protocol integrating qualitative and quantitative components. Biomarkers 3:109–127.

Köhler H-R, Triebskorn R, Stöcker W, Kloetzel P-M, Alberti G (1992) The 70 kD heat shock protein in soil invertebrates: a possible tool for monitoring environmental contaminants. Arch Environ Contam Toxicol 22:334–338.

Kratz W (1998) The Bait-Lamina test, general aspects, applications and perspectives. Environ Sci Pollut Res 5:94–96.

Lagadic L, Caquet T, Ramade F (1994) The role of biomarkers in environmental assessment. (5) Invertebrate populations and communities. Ecotoxicology 3:180–192.

Lakhani L, Bhatnagar BS, Pandey AK (1991) Effect of monocrotophos on the ovary of the earthworm *Eudichogaster kinneari* (Stephenson). A histological and histochemical profile. J Reprod Biol Comp Endocrinol 3:39–46.

Langenscheidt M (1973) Zur Wirkungsweise von Sterilität erzeugenden Stoffen bei *Tetranychus urticae* Koch (Acari, Tetranychidae). II. Wirkungsweise von Apholate bei Spinnmilben. Z Angew Entomol 74:142–151.

Langenscheidt M (1975) Zur Wirkungsweise von Sterilität erzeugenden Stoffen bei *Tetranychus urticae* Koch (Acari, Tetranychidae). III. Wirkungsweise von 5-Fluoruracil bei Spinnmilben. Z Angew Entomol 78:160–170.

Langer T, Neupert W (1990) Heat shock proteins hsp60 and hsp70: their roles in folding, assembly and membrane translocation of proteins. In: Kaufmann SHE (ed) Heat Shock Proteins and Immune Response. Current Topics in Microbiology and Immunology, Vol. 167. Springer-Verlag, Berlin.

Large AT, Connock MJ (1994) Centrifugal evidence for association of mannitol oxidase with distinct organells ("mannosomes") in the digestive gland of several species of terrestrial gastropod mollusc. Comp Biochem Physiol A 107:621–629.

Lastowski-Perry D, Otto E, Maroni G (1985) Nucleotide sequence and expression of a *Drosophila* metallothionein. J Biol Chem 260:1527–1530.

Lee HC, Aarhus R (1993). Wide distribution of an enzyme that catalyses the hydrolysis of cyclic ADP-ribose. Biochem Biophys Acta 1164:68–74.

Løkke H, Van Gestel CAM (eds) (1998) Handbook of Soil Invertebrate Toxicity Tests. Wiley, Chichester.

Lowe DM, Moore MN, Clarke KR (1981) Effects of oil on digestive cells in mussels: quantitative alterations in cellular and lysosomal structure. Aquat Toxicol 1:213–226.

Ludwig M, Kratzmann M, Alberti G (1991) Accumulation of heavy metals in two oribatid mites. In: Dussbek F, Bukva V (eds) Modern Acarology, vol 1. Academia, Prague, and SPB Academic Publishing bv, The Hague, pp 431–437.

Manchenko GP (1994) Detection of Enzymes on Electrophoretic Gels: A Handbook. CRC Press, Boca Raton, FL.

Margoshes M, Vallee BL (1957) A cadmium protein from equine cortex. J Am Chem Soc 79:4813–4818.
Marigómez I, Soto M, Kortabitate M (1996) Tissue-level biomarkers and biological effect of mercury on sentinel slugs, *Arion ater*. Arch Environ Contam Toxicol 31: 54–62.
Marigómez JA, Cajaraville MP, Angulo E, Moja J (1990) Ultrastructural alterations in the renal epithelium of cadmium treated *Littorina littorea* (L.). Arch Environm Contam Toxicol 19:863–881.
Martin J, Horwich AL, Hartl FU (1992) Prevention of protein denaturation under heat stress by the chaperonin HSP60. Science 258:995–998.
Mayer FL, Versteeg DJ, McKee MJ, Folmar LC, Graney RL, McCume DC, Rattner BR (1992) Physiological and nonspecific biomarkers. In: Huggett, RJ, Kimerle, RA, Mehrle Jr. PM, Bergman HL (eds) Biomarkers. Biochemical, Physiological, and Histological Markers of Anthropogenic Stress. Lewis, Boca Raton, FL, pp 5–85.
McCluskey S, Mather PB, Hughes JM (1993) The relationship between behavioural responses to temperature and genotype at a PGI locus in the terrestrial isopod *Porcellio laevis*. Biochem Syst Ecol 21:171–179.
McDonald IC, Krysan JL, Johnson OA (1990) Studies of electrophoretic variation in *Diabrotica* as influenced by the age, sex, or diet of adult beetles (Coleoptera: Chrysomelidae). Ann Entomol Soc Am 83:1192–1202.
McMullin TW, Halberg RL (1987) A normal mitochondrial protein selectively synthesized and accumulated during heat shock in *Tetrahymena termophila*. Mol Cell Biol 7:4414–4423.
Migula P (1999) Enzymatic effects and animal population demography in polluted environments. In: Kammenga, JE, Laskowski R (eds) Demographic Ecotoxicology. Wiley, London, in press.
Moore MN (1985) Cellular responses to pollutants. Mar Poll Bull 16:134–139.
Moore MN (1988) Cellular and histopathological effects of a pollutant gradient—introduction. Mar Ecol Prog Ser 46:79.
Moore MN (1990). Lysosomal cytochemistry in marine environmental monitoring. Histochemistry 22:187–191.
Moradeshagh MJ, Brindley WA, Youssef NN (1974) Chlorcyclizine and SKF 525A effects on parathion toxicity and midgut tissue structure in alkali bees, *Nomia melandesi*. Environ Entomol 3:455–463.
Morimoto R, Abravaya K, Mosser G, Williams GT (1990) Transcription of the human hsp70 gene: cis-acting elements and trans-acting factors involved in basal, adenovirus E1A, and stress-induced expression. In: Schlesinger MJ, Santoro MG, Garaci E (eds) Stress Proteins. Induction and Function. Springer-Verlag, New York, pp 1–17.
Mothes U (1981) Feinstrukturelle Veränderung am Integument von *Tetranychus urticae* Koch (Acari, Tetranychidae) nach Nikkomycin-Behandlung (AMS 0896 Bayer Leverkusen). Mitt Dtsch Ges Allg Angew Entomol 2:172–179.
Mothes U, Seitz K-A (1982) Action of the microbial metabolite and chitin synthesis inhibitor Nikkomycin on the mite *Tetranychus urticae*; an electron microscope study. Pestic Sci 13:426–441.
Mothes-Wagner U, Reitze HK, Seitz K-A (1990) Environmental actions of agrochemicals. 2. Histological effects of the herbicide/insecticide dinoseb-acetate (2-sec-butyl-4,6-dinitrophenyl acetate) on the spider mite *Tetranychus urticae* (Acari: Tetranychidae) reared on herbicide-treated *Phaseolus vulgaris*. Exp Appl Acarol 9:289–310.

Mutwakil MHAZ, Reader JP, Holdich DM, Smithurst PR, Candido EPM, Jones D, Stringham EG, de Pomerai DI (1997) Use of stress-inducible transgenic nematodes as biomarkers of heavy metal pollution in water samples from an English river system. Arch Environ Contam Toxicol 32:146–153.

Neumann W (1985) Veränderungen am Mitteldarm von *Oxidus gracilis* (C.L. Koch, 1847) während einer Häutung (Diplopoda). Bijdr Dierkd 55:149–158.

Nicholson JK, Timbrell J, Sadler PJ (1985). Proton NMR spectra of urine as indicators of renal damage—mercury induced nephrotoxicity in rats. Mol Pharm 27:644–651.

Nicholson JK, Wilson ID (1989) High resolution proton magnetic resonance spectroscopy of biological fluids. Prog NMR Spectros 21:449–501.

Nover L (1984) Heat Shock Response in Eucaryotic Cells. Springer-Verlag, New York, pp 1–82.

Øien N, Stenersen J (1984) Esterases of earthworms. III. Electrophoresis reveals that *Eisenia fetida* (Savigny) is two species. Comp Biochem Physiol 78:277–282.

Olesen TME (1998) The use of an earthworm biomarker to assess the effects of sewage sludge applications to agricultural land. MSc thesis, University of Odense.

Orbea A, Marigomez I, Fernandez C, Tarazona JV, Cancio I, Cajaraville MP (1999) Structure of peroxisomes and activity of the marker enzyme catalase in digestive epithelial cells in relation to PAH content of mussels from two Basque estuaries (Bay of Biscay): seasonal and site-specific variations. Arch Environ Contam Toxicol 36: 158–166.

Pasteur N, Iseki A, Georghiou GP (1981) Genetic and biochemical studies of the highly active esterases A' and B associated with organophosphate resistance in mosquitoes of the *Culex pipiens* complex. Biochem Genet 19:909–919.

Pawert M, Müller E, Triebskorn R (1998) Ultrastructural changes in fish gills as biomarkers to assess small stream pollution. Tissue Cell 30:617–626.

Pawert M, Rahmann H, Köhler H-R (1994) The impact of heavy metals on the ultrastructure of the endoplasmic reticulum in midgut epithelial cells of *Tetrodontophora bielanensis* (Collembola). Verh Dtsch Zool Ges 87:331.

Pawert M, Triebskorn R, Graff S, Berkus M, Schulz J, Koehler H-R (1996) Cellular alterations in collembolan midgut cells as a marker of heavy metal exposure: ultrastructure and intracellular metal distribution. Sci Total Environ 181:187–200.

Peakall DB (1994) The role of biomarkers in environmental assessment 1. Introduction. Ecotoxicology 3:157–160.

Peakall DB, Shugart LR (eds) (1992) Strategy for Biomarker Research and Application in the Assessment of Environmental Health. Springer-Verlag, Heidelberg.

Peakall DB, Walker CH (1994) The role of biomarkers in environmental assessment. (3) Vertebrates. Ecotoxicology 3:173–179.

Pedersen SN, Lundebye AK, Depledge MH (1997) Field application of metallothionein and stress protein biomarkers in the shore crab (*Carcinus maenas*) exposed to trace metals. Aquat Toxicol 37:183–200.

Peiris HTR, Hemingway J (1993) Characterization and inheritance of elevated esterases in organophosphorus and carbamate insecticide resistant *Culex quinquefasciatus* (Diptera: Culicidae) from Sri Lanka. Bull Entomol Res 83:127–132.

Pelham, HRB (1986) Speculations on the function of the major heat shock and glucose-regulated proteins. Cell 46:959–961.

Perdew GH (1988) Association of the Ah receptor with the 99 kD heat shock protein. J Biol Chem 263:13802–13805.

Plesofsky-Vig N (1996) The heat shock proteins and the stress response. In: Marzluf B (ed) The Mycota, vol III . Springer-Verlag, pp 171–190.

Prabhakaran SK, Kamble ST (1993) Activity and electrophoretic characterization of esterases in insecticide-resistant and susceptible strains of German cockroach (Dictyoptera: Blattellidae). J Econ Entomol 86:1009–1013.

Prosi F, Dallinger R (1988) Heavy metals in the terrestrial isopod *Porcellio scaber* Latreille. I. Histochemical and ultrastructural characterization of metal-containing lysosomes. Cell Biol Toxicol 4:81–96.

Prosi F, Storch V, Janssen HH (1983) Small cells in the midgut glands of terrestrial Isopoda: sites of heavy metal accumulation. Zoomorphology 102:53–64.

Proux J, Alaoui A, Moretau B, Baskali A (1993) Deltamethrin-induced deregulation of the water balance in the migratory locust *Locusta migratoria*. Comp Biochem Physiol C 106:351–357.

Pyza E, Mak P, Kramarz P, Laskowski R (1997) Heat shock proteins (HSP70) as biomarkers in ecotoxicological studies. Ecotoxicol Environ Saf 38:244–251.

Rankevich D, Lavie B, Nevo E, Beiles A, Arad Z (1996) Genetic and physiological adaptations of the prosobranch landsnail *Pomatias olivieri* to microclimatic stresses on Mount Carmel, Israel. Isr J Zool 42:425–441.

Raymond M, Callaghan A, Fort P, Pasteur N (1991) Worldwide migration of insecticide resistance genes in mosquitoes. Nature (Lond) 350:151–153.

Raymond M, Fournier D, Bride J-M, Cuany A, Berge J, Magnin M, Pasteur N (1986) Identification of resistance mechanisms in *Culex pipiens* (Diptera: Culicidae) from Southern France. Insensitive acetylcholinesterase and detoxifying oxidases. J Econ Entomol 79:1452–1458.

Raymond M, Pasteur N, Georghiou G, Mellon RB, Wirth MC, Hawley MK (1987) Detoxification esterases new to California, USA, in organophosphate-resistant *Culex quinquefasciatus* (Diptera: Culicidae). J Med Entomol 24:24–27.

Raymond M, Poulin E, Boiroux V, Dupont E, Pasteur N (1993) Stability of insecticide resistance due to amplification of esterase genes in *Culex pipiens*. Heredity 70:301–307.

Recio A, Marigomez JA, Angulo E, Moya J (1988) Zinc treatment of the digestive gland of the slug *Arion ater* L. 1. Cellular distribution of zinc and calcium. Bull Environ Contam Toxicol 41:858–864.

Reich T, Depew MC, Marks GS, Singer MA, Wan JKS (1981) Effect of polychlorinated biphenyls on phospholipid membrane fluidity. J Environ Sci Health A Environ Sci Eng 16:65–72.

Réz G (1986) Electron microscopic approaches to environmental toxicity. Acta Biol Hung 37(1):31–45.

Riandry M-F (1993) Carboxylic esterases sytem in *Anopheles stephensi*. C R Acad Sci Paris 316:1183–1187.

Ridlington JW, Chapman DC, Goeger DE, Whanger PD (1981) Metallothionein and Cuchelatin: characterization of metal-binding proteins from tissues of four marine animals. Comp Biochem Physiol B 70:93–104.

Rizvi SSA, Khan MA (1973) Histopathological effects of certain insecticides on the malpighian tubules of *Hieroglyphus nigrorepletus* Bol. (Acrididae:Orthoptera). Z Angew Entomol 73:400–405.

Rumpf S, Storch V, Vogt H, Hassan SA (1992) Effects of juvenoids on larvae of *Chrysoperla carnea* (Chrysopidae). Acta Phytopathol Entomol Hung 27:557–563.

Sanders BM (1993) Stress proteins in aquatic organisms: an environmental perspective. Crit Rev Toxicol 23:49–75.

Sanders BM, Dyer SD (1994) Cellular stress response. Environ Toxicol Chem 13:1209–1210.

Sanders BM, Martin LS, Nakagawa PA, Hunter DA, Miller S, Ullrich SJ (1994) Specific cross-reactivity of antibodies raised against two major stress proteins, stress 70 and chaperonin 60 in diverse species. Environ Toxicol Chem 13:1241–1249.

Sato M, Sasaki M, Hojo H (1993) Induction of metallothionein synthesis by oxidative stress and possible role in acute phase response. In: Suzuki KT, Imura N, Kimura M, (eds) Metallothionein III. Birkhäuser, Basel, pp 29–55.

Schlesinger MJ, Ashburner M, Tissiéres A (1982) Heat Shock—from Bacteria to Man. Cold Spring Harbor Laboratory, Cold Spring Harbor, New York.

Schlesinger MJ, Santoro MG, Garaci E (1990) Stress Proteins. Induction and Function. Springer-Verlag, New York.

Schlüter U (1984) Disturbance of epidermal and fat body tissue after feeding Azadirachtin and its consequence on larval moulting in the Mexican bean beetle *Epilachna varivestis* (Coleoptera: Coccinellidae). Entomol Sener 10:97–110.

Schlüter U, Seifert G (1988) Inducing melanotic nodules within the fat body of the last instar larvae of *Epilachna varivestis* (Coleoptera) by Azadirachtin. J Invert Pathol 51:1–9.

Schramm M, Müller E, Triebskorn R (1998) Brown trout (*Salmo trutta* f. Fario) liver ultrastructure as a biomarker for assessment of small stream pollution. Biomarkers 3:93–108.

Segner H, Braunbeck T (1990) Adaptive changes of liver composition and structure in golden ide during winter acclimatization. J Exp Zool 255:171–185.

Segner H, Juario JV (1986) Histological observations on the rearing of milkfish *Chanos chanos* fry using different diets. J Appl Ichthyol 2:162–173.

Slice LW, Freedman JH, Rubin C (1990) Purification, characterization, and cDNA cloning of a novel metallothionein-like, cadmium-binding protein from *Caenorhabditis elegans*. J Biol Chem 265:256–263.

Somlyo AP, Garfield RE, Chacko S, Somlyo AV (1975) Golgi organelle response to the antibiotic X 537 A. J Cell Biol 66:425–443.

Sorger PK, Nelson H (1989) Trimerization of a yeast transcriptional activator via a coiled-cell-motif. Cell 59:807–813.

Sparks AK (1972) Invertebrate Pathology, Noncommmunicable Diseases. Academic Press, New York.

Sparks AK (1985) Synopsis of Invertebrate Pathology. Elsevier, Amsterdam.

Stegeman JJ, Brouwer M, Di Giulio RT, Förlin L, Fowler BA, Sanders BM, Van Veld PA (1992) Molecular responses to environmental contamination: enzyme and protein synthesis as indicators of chemical exposure and effect. In: Huggett RJ, Kimerle RA, Mehrle PM Jr, Bergman HL (eds) Biomarkers: Biochemical, Physiological, and Histological Markers of Anthropogenic Stress. Lewis, Boca Raton, FL, pp 235–335.

Storch V (1988) Cell and environment: a link between morphology and ecology. In: Iturrondobeitia JC (ed) Biologia Ambiental I. Serv Ed Univers Pais Vasco 1:179–191.

Stringham EG, Candido EPM (1993) Targeted single-cell induction of gene products in *Caenorhabditis elegans*: a new tool for developmental studies. J Exp Zool 266:227–223.

Stringham EG, Candido EPM (1994) Transgenic hsp16-lacZ strains of the soil nematode

Caenorhabditis elegans as biological monitors of environmental stress. Environ Toxicol Chem 13:1211–1220.

Stringham EG, Dixon DK, Jones D, Candido EPM (1992) Temporal and spatial expression patterns of the small heat shock (hsp16) genes in transgenic *Caenorhabditis elegans*. Mol Cell Biol 3:221–233.

Štrus J, Burkhardt P, Storch V (1985) The ultrastructure of the midgut glands in *Ligia italica* (Isopoda) under different nutritional conditions. Helgol Meeresunters 39:367–374.

Studer R, Vogt CP, Gavicelli M, Hunziker PE, Kägi JHR (1997) Hepatic metallothionein in human hepatic cells is linked to cellular proliferation. Biochem J 328:63–97.

Stürzenbaum SR, Kille P, Morgan AJ (1998) The identification, cloning and characterization of earthworm metallothionein. FEBS Lett 431:437–442.

Suzuki K, Hama H, Konno Y (1993) Carboxylesterase of the cotton aphid, *Aphis gossypii* Glover (Homoptera: Aphididae), responsible for fenitrothion resistance as a sequestering protein. Appl Entomol Zool 28:439–450.

Suzuki KT, Yamamura M, Mori T (1980) Cadmium-binding proteins induced in the earthworm. Arch Environ Contam Toxicol 9:415–424.

Svendsen C, Meharg AA, Freestone P, Weeks JM (1996) Use of an earthworm lysosomal biomarker for the ecological assessment of pollution from an industrial plastics fire. Appl Soil Ecol 3:99–107.

Svendsen C, Weeks JM (1995) The use of a lysosome assay for the rapid assessment of cellular stress from copper to the freshwater snail *Viviparus contectus* (Millet). Mar Pollut Bull 31(1–3):139–142 (special issue).

Svendsen C, Weeks JM (1997a) Relevance and applicability of a simple earthworm biomarker of copper exposure: I. Links to ecological effects in a laboratory study with *Eisenia andrei*. Ecotoxicol Environ Saf 36:72–79.

Svendsen C, Weeks JM (1997b) Relevance and applicability of a simple earthworm biomarker of copper exposure: II Validation and applicability under field conditions in mesocosm experiment with *Lumbricus rubellus*. Ecotoxicol Environ Saf 36:80–88.

Svendsen C, Spurgeon DJ, Zvezdelin BM, Weeks JM (1998) Lysosomal membrane permeability and earthworm immune-system activity; field-testing on contaminated land. In: SETAC Press, Sheppard S, Reinecke AJ, Posthuma L, Holmstrup M (eds) Advances in Earthworm Ecotoxicology. Pensacola, FL, pp 225–232.

Sylvie BR, Pairault C, Vernet, Boulebache H (1996) Effect of lindane on the ultrastructure of the liver of the rainbow trout, *Oncorhynchus mykiss*, sac-fry. Chemosphere 33:2065–2079.

Tedesco JL, Courtright JB, Kumaran AK (1986) Ultrastructural changes induced by juvenile hormone analogue in oocyte membranes of apterous *Drosophila melanogaster*. J Insect Physiol 27:895–902.

Theodorakis CW, D'Surney SJ, Bickham JM, Lyne TB, Bradley BP, Hawkins WE, Farkas WL, McCarthy JF, Shugart LR (1992) Sequential expression of biomarkers in bleugill sunfish exposed to contaminated sediments. Ecotoxicology 1:45–73.

Thomas JP, Bachowski GJ, Girotti AW (1986) Inhibition of cell membrane lipid peroxidation by cadmium- and zinc-metallothionein. Biochim Biophys Acta 884:448–461.

Thompson HM, Walker CH, Hardy AR (1988) Avian esterases as indicators of exposure to insecticides–the factor of diurnal variation. Bull Environ Contam Toxicol 41:4–11.

Thornalley PJ, Vašák M (1985) Possible role for metallothionein in protection against radiation-induced oxidative stress. Kinetics and mechanism of its reaction with superoxide and hydroxyl radicals. Biochim Biophys Acta 827:36–44.

Tissiéres A, Mitchell HK, Tracy UM (1974) Protein synthesis in the salivary glands of *D. melanogaster*. Relation to chromosome puffs. J Mol Biol 84:389–398.

Tranvik L, Sjögren M, Bengtsson G (1994) Allozyme polymorphism and protein profile in *Orchesella bifasciata* (Collembola). Indicative of extended metal pollution? Biochem Syst Ecol 22:13–23.

Triebskorn R (1989) Ultrastructural changes in the digestive tract of *Deroceras reticulatum* (Müller) induced by a carbamate molluscicide and by metaldehyde. Malacologia 31:141–156.

Triebskorn R (1991) Cytological changes in the digestive system of slugs induced by molluscicides. J Med Appl Malacol 3:113–123.

Triebskorn R (1995) Tracing of molluscicides and cellular reactions induced by them in slugs' tissues. In: Cajaraville MP (ed) Cell Biology in Environmental Toxicology. pp 193–220.

Triebskorn R, Christensen K, Heim I (1998) Effects of orally and dermally applied metaldehyde on mucus cells of slugs (*Deroceras reticulatum*) depending on temperature and duration of exposure. J Moll Stud 64:467–487.

Triebskorn R, Ebert D (1989) The importance of mucus production in slugs' reaction to molluscicides and the impact of molluscicides on the mucus producing system. Proc Br Crop Prot Counc 41:373–379.

Triebskorn R, Henderson IF, Martin A (1999) Detection of iron in tissues from slugs (*Deroceras reticulatum*) (Müller) after ingestion of iron chlates, by means of energy-filtering transmission electron microscopy (EFTEM). Pestic Sci 55:55–61.

Triebskorn R, Henderson IF, Martin A, Köhler H-R (1996) Slugs as target and non-target organisms for environmental pollution. Br Crop Prot Counc 66:65–72.

Triebskorn R, Köhler H-R (1992) Plasticity of the endoplasmatic reticulum in three cell types of slugs poisoned by molluscicides. Protoplasma 169:120–129.

Triebskorn R, Köhler H-R (1996) The impact of heavy metals on the grey garden slug *Deroceras reticulatum* (Müller): Metal storage, cellular effects and semi-quantitative evaluation of metal toxicity. Environ Pollut 93:327–343.

Triebskorn R, Köhler H-R, Honnen W, Schramm M, Adams SM, Mueller EF (1997) Induction of heat shock proteins, changes in liver ultrastructure, and alterations of fish behavior: Are these biomarkers related and are they useful to reflect the state of pollution in the field? J Aquat Ecosyst Stress Recov 6:57–73.

Triebskorn R, Köhler H-R, Zahn T, Vogt G, Ludwig M, Rumpf S, Kratzmann M, Alberti G, Storch V (1991) Invertebrate cells as targets for hazardous substances. Z Angew Entomol 78:277–287.

Triebskorn R, Künast C (1990) Ultrastructural changes in the digestive system of *Deroceras reticulatum* (Mollusca; Gastropoda) induced by lethal and sublethal concentrations of the carbamate molluscicide Cloethocarb. Malacologia 32:89–106.

Triebskorn R, Künast C, Huber R, Brem G (1989) The tracing of a ^{14}C-labeled carbamate molluscicide through the digestive system of *Deroceras reticulatum*. Pestic Sci 28: 321–330.

Van Gestel CAM, Van Brummelen TC (1996) Incorporation of the biomarker concept in ecotoxicology calls for a redefinition of terms. Ecotoxicology 5:217–225.

Vedel GR, Depledge MH (1995) Stress-70 levels in the gills of *Carcinus maenas* exposed to copper. Mar Pollut Bull 31(1–3):84–86 (special issue).

Vijayan MM, Pereira C, Forsyth RB, Kennedy CJ, Iwama GK (1997) Handling stress does not affect the expression of hepatic heat shock protein 70 and conjugation enzymes in rainbow trout treated with beta-naphthoflavone. Life Sci 61:117–127.

Vincent M, Tanguay RM (1982) Different intracellular distributions of heat-shock and arsenite-induced proteins in *Drosophila* Kc cells. J Mol Biol 162:365–378.

Vogt G (1991) In vivo decondensation of chromatin and nucleolar fibrillar component by *Leucaena leucocephala* ingredient. Biol Cell 72:211–215.

Vogt G, Böhm R, Segner H (1994) Mimosine-induced cell death and related chromatin changes. J Submicrosc Cytol Pathol 26:319–330.

Walker CH (1995) Biochemical biomarkers in ecotoxicology–Some recent developments. Sci Total Environ 171:189–195.

Weeks JM (1998) Effects of pollutants on soil invertebrates: links between levels. In: Schüürmann G, Markert, B (eds) Ecotoxicology, Ecological Fundamentals, Chemical Exposure and Biological Effects. Wiley, New York, pp 645–662.

Weeks JM, Evdokimova G, Ernst WHO, Scott-Fordsmand J, Sousa JP, Sergeev VE, Nakonieczny M (1999) Strategy for the implementation of a biomarker biomonitoring programme in the Kola Peninsula, Russia. In: Peakall DB, Walker CH, Migula P (eds) Proceedings of the NATO Advanced Research Workshop on Biomarkers: A pragmatic basis for remediation of severe pollution in Europe. NATO Science Series 2, Environmental Security, vol 54.

Weeks JM, Svendsen C (1996) Neutral-red retention by lysosomes from earthworm (*Lumbricus rubellus*) coelomocytes: a simple biomarker of exposure to soil copper. Environ Toxicol Chem 15:1801–1805.

Westheide W, Bethke-Beilfuß D, Hagens M, Brockmeyer V (1989) Enchytraeiden als Testorganismen—Voraussetzungen fuhr ein terrestrisches Testverfahren und Testergebnisse. Verh Ges Ökol 17:793–798.

Willuhn J, Schmitt-Wrede HP, Greven H, Wunderlich F (1994) cDNA cloning of a cadmium-inducible mRNA encoding a novel cysteine-rich, non-metallothionein 25-kDa protein in an enchytraeid earthworm. J Biol Chem 269:24688–24691.

Wirth MC, Marquine M, Georghiou GP, Pasteur N (1990) Esterases A2 and B2 in *Culex quinquefasciatus* (Diptera: ulicidae): role in organophosphate resistance and linkage. J Med Entomol 27:202–206.

Zanger M, Köhler H-R (1996) Colour change: a novel biomarker indicating sublethal stress conditions in the millipede *Julus scandinavius* (Diplopoda). Biomarkers 1:99–106.

Zanger M, Harreus D, Alberti G, Köhler H-R (1994) Different methods for qualifying and quantifying hsp 70 in diplopods. Verh Dtsch Zool Ges 87:335.

Zanger M, Alberti G, Kuhn M, Köhler H-R (1996) The stress-70 (hsp70) protein family in diplopods: induction and characterization. J Comp Physiol B 165:622–627.

Zeng J, Heuchel R, Schaffner W, Kägi JHR (1991) Thionein (apometallothionein) can modulate DNA binding and transcription activation by zinc finger containing factor Sp1. FEBS Lett 279:310–312.

Manuscript received June 3, 1999; accepted June 12, 1999.

Index

Abamectin, household insecticide, 34
Allethrin, household insecticide, 33
Analytical methods, metalaxyl, 1 ff., 9
Aquatic mercury, DOC related, 78, 84
Attractants, household pesticides, 35

Bendiocarb, household insecticide, 32
Bioconcentration, environmental mercury, 79
Biomagnification, environmental mercury, 79
Biomarker continuum (illus.), 95
Biomarkers (Cadmium) in *Helix pomatia* (diag.), 107
Biomarkers, defined, 94
Biomarkers, ecotoxicological soil risk assment, 93 ff.
Biomarkers, novel, 118
Biomarkers of effect, defined, 95
Biomarkers of exposure, defined, 95
Biomarkers of susceptibility, defined, 95
Biomarkers, terrestrial invertebrates, 93 ff.
Biomonitoring, soil invertebrate biomarkers, 126
BIOPRINT project, invertebrate biomarker detection, 96
Boric acid, household insecticide, 34
Brodifacoum, household rodenticide, 36
Bromadiolone, household rodenticide, 36

Cancer risk, lifetime pesticide contaminated drinking water, 56
Capillary electrophoresis, fungicide in water, 14
Carbamate household insecticides, 32
Carbaryl, household insecticide, 32
Carcinogenicity, household pesticides, 50, 53
CAS Registry numbers, household pesticides, 58
Cd biomarkers in *Helix pomatia* (diag.), 107
Cedarwood oil, household insecticide, 36

Chaperonin stress protein, 100
Chemical names, household pesticides, 58
Chlorophacinone, household rodenticide, 36
Chlorpyrifos, household insecticide, 31
Cholecalciferol, household rodenticide, 36
Citronella oil, household insecticide, 35
Common names, household pesticides, 58
Cyfluthrin, household insecticide, 34
Cypermethrin, household insecticide, 33

Deet, household insecticide, 35
Deltamethrin, household insecticide, 33
Dermal LD_{50}s, pesticide technical materials, 50
Diazinon, household insecticide, 31
p-Dichlorobenzene, household insecticide, 36
Dichlorvos, household insecticide, 31
Dilution attenuation factor, pesticides landfills, 48
Dimethylmercury, atmospheric photo-converted, 72
Diphacinone, household rodenticide, 36
Disinfectants (household), EPA categories, 43H
Disposal costs, hazardous household waste, 28
Disposal of household pesticides, risk assessment, 46
Dissolved organic carbon (DOC), mercury complexing ligand, 77
DOC (dissolved organic carbon), aquatic mercury, 78, 84

Early warning system, transgenic invertebrates, 102
Ecotoxicological soil risk assessment, 93 ff.
Electron microscopy, diagnostic histology tool, 110
ELISA, metalaxyl analysis, 15
Environmental mercury, 70
EPACMTP, pesticide leaching model, 47

Esbiothrin, household insecticide, 33
Eserine-sensitive esterases, earthworm biomarker, 114
Esfenvalerate, household insecticide, 34

Fenoxycarb, household insecticide, 34
Fenvalerate, household insecticide, 33
Fepronil, household insecticide, 35

Gas liquid chromatography, metalaxyl, 11
Gel electrophoresis, isozyme separation, 114
Golgi apparatus, biomarker, 112
Groundwater, household pesticides detected, 40
Groundwater, pesticide leaching criteria, 39

Hazard quotient values, household pesticides, 54
Hazard quotients, pesticide contaminated water, 54
Hazardous household waste, disposal costs, 28
Hazardous household waste disposal, options 30
Hazardous household waste, examples, 29
Heat shock proteins, invertebrate biomarkers, 97
Hepatopancreas, gastropod biomarker organ, 111
Hg^{+2} (particulate form), 83
Hg^0 (metallic form), 71, 83
Hg^{2+} (mercuric form), 71, 83
Hg_2^{2+} (mercurous form), 71, 83
High performance liquid chromatography (HPLC), metalaxyl, 14
Histology, invertebrate as biomarker, 109
Household disinfectants, EPA categories, 43
Household hazardous waste, 27, 29
Household hazardous waste disposal, options, 30
Household pesticide hazardous waste, 27 ff.
Household pesticides, 27 ff.
Household pesticides, amounts disposed landfills, 45
Household pesticides, CAS Registry numbers, 31, 58

Household pesticides, chemical names, 58
Household pesticides, common names list, 58
Household pesticides, detected groundwater, 40
Household pesticides, disposal risks, 46
Household pesticides, exceeding leaching criteria, 39
Household pesticides, found in groundwater, 40
Household pesticides, hazard quotient values, 54
Household pesticides, indoor defined, 28
Household pesticides, inhalation LC_{50}s, 50
Household pesticides, K_{oc} values, 31
Household pesticides, LD_{50}s (dermal/oral), 50
Household pesticides, leaching potential landfills, 38
Household pesticides, Log K_{ow} values, 31
Household pesticides, physical/chemical properties, 31
Household pesticides, Q^*_1 values, 50
Household pesticides, soil half-lives, 31
Household pesticides, toxicity of technical materials, 50
HPLC, metalaxyl, 14
Human health risk, household pesticides disposal, 46
Hydramethylnon, household insecticide, 34
Hydrolysis, metalaxyl, 5
Hydroprene, household insecticide, 34

In-lake mercury methylation, 84
In-lake methylation, major source methylmercury, 84
Incineration, household pesticide waste option, 37
Indoor household pesticides, 27 ff.
Indoor household pesticides, common names list, 31
Indoor household pesticides, defined, 28
Inhalation LC_{50}s, pesticide technical materials, 50
Insect attractants, household pesticides, 35
Invertebrate biomarker research, European Union, 96
Invertebrate biomarkers, condfounding factors, 120

Invertebrate biomarkers, ecological relevance, 125
Invertebrate histology, as biomarker, 109
Invertebrate ultrastructure, as biomaker, 109
Isozyme separation, gel electrophoresis, 114
Isozymes, biomarker ChE inhibitor exposure, 115
Isozymes, invertebrate biomarkers, 113

K_{oc}s, household pesticides, 31
K_{ow}s, household pesticides, 31

Landfills, improved standards, 57
Landfills, number in U.S., 37
Lavandin oil, household insecticide, 36
Leaching potential, household pesticides, 39
Lifetime cancer risk, household pesticides, 48
Lifetime cancer risk, pesticides drinking water, 56
Limonene, household insecticide, 35
Linalool, household insecticide, 35
Log K_{ow}s, household pesticides, 31
Log P_{ow}, metalaxyl, 16
Low molecular weight stress proteins, 101
Lysosomal membrane integrity, invertebrate biomarker, 115
Lysozyme activity, biomarker earthworm metal immunotoxicity, 116

Mass spectrometry, metalaxyl, 15
Maximum residue limits (MRL), metalaxyl, 2
Mercury contamination, atmospheric transport, 74
Mercury contamination/bioaccumulation aquatic systems, 69 ff.
Mercury cycling (diag.), 73, 74
Mercury cycling model (illus.), 83
Mercury cycling models, existing, 82
Mercury, environmental biomagnification, 79
Mercury, environmental sources, 73
Mercury, in-lake methylation, 84
Mercury methylation-demethylation, pH related, 78

Mercury modeling, aquatic ecosystems, 69 ff.
Mercury modeling, bioaccumulation aquatic systems, 69 ff.
Mercury, modeling in aquatic systems, 80
Mercury poisoning, Minamata Bay (Japan), 70
Mercury, sedimentation major aquatic sink, 85
Mercury speciation, 70
Mercury, species classification, 71
Mercury, various vapor pressures, 73, 82
Metal-binding proteins, invertebrate biomarkers, 104
Metalaxyl, abiotic degradation, 4
Metalaxyl, ADI, 4
Metalaxyl, analytical methods, 1 ff., 9
Metalaxyl, bioassay in soil, 15
Metalaxyl, biotic degradation, 5
Metalaxyl, chemical physical properties, 3
Metalaxyl, chemical structure, 3
Metalaxyl, degradation, 1 ff.
Metalaxyl, ELISA analysis, 15
Metalaxyl fungicide, 1 ff.
Metalaxyl, gas liquid chromatography analysis, 11
Metalaxyl, HPLC analysis, 14
Metalaxyl, hydrolysis, 5
Metalaxyl, log $P_{ow,}$ 16
Metalaxyl, mass spectrometric analysis, 15
Metalaxyl, maximum residue limits (MRL), 4
Metalaxyl, metabolic pathway in rat, 9, 10
Metalaxyl, metabolism, 1 ff., 6
Metalaxyl, metabolism in plants, 8
Metalaxyl, metabolism in soil, 6, 8
Metalaxyl, micellar electrokinetic capillary chromatography analysis, 14
Metalaxyl, mode of action, 2
Metalaxyl, NOEL, 4
Metalaxyl, persistence/degradation, 1 ff
Metalaxyl, photodegradation, 5, 7
Metalaxyl, photoproducts, 7
Metalaxyl, plant diseases controlled, 1
Metalaxyl, plant uptake/translocation, 16
Metalaxyl, S-enantiomer form, 2
Metalaxyl, soil adsorption/mobility, 18
Metalaxyl, soil breakdown, 8

Metalaxyl, soil-adsorbed via hydrogen bonding, 18
Metalaxyl, systemic plant fungicide, 2
Metalaxyl, thin layer chromatography analysis, 11
Metalaxyl, toxicity, 3
Metalaxyl, translocation/persistence plants, 16
Metallothionein isoforms, *Helix pomatia*, 107
Metallothioneins, earthworms, 106
Metallothioneins, invertebrate biomarkers, 104
Metallothioneins, snails, 105
Methomyl, household insecticide, 32
Methooprene, household insecticide, 34
Methylmercury, bacteria-produced, 76
Methylmercury, in-lake methylation major source, 84
Methylmercury, natural production, 72
MGK-264, household insecticide, 33
MGK-326, household insecticide, 35
Micellar electrokinetic capillary chromatography, metalaxyl, 14
Minamata Bay (Japan), chronic mercury poisoning, 70
Mode of action, metalaxyl, 2
Modeling, mercury in aquatic systems, 80
Moth proofers, household pesticides, 35
MRL, see Maximum residue limits, 4

Naled, household insecticide, 31
Naphthalene, household insecticide, 35
Neopynamin, household insecticide, 33
Neutral red retention vs. body copper levels, invertebrates, 117
NMR spectroscopy, toxicological biomarker research, 118
Novel biomarkers, 118

Oral cancer slope factor (Q^*_1), household pesticides, 47, 50
Oral LD_{50}s, pesticide technical materials, 50
Organism age vs. methylmercury content, 79
Organophosphate household insecticides, 31

Paradichlorobenzene, household insecticide, 36
Patchouli oil, household insecticide, 35
Pennyroyal, household insecticide, 35
Permethrin, household insecticide, 33
Pesticide chemical names, 58
Pesticide leaching model (EPACMTP), 47
Pesticides, exceeding leaching criteria, 39
Pesticides (household), amounts disposed landfills, 45
Pesticides (household), formulation numbers, 30
Petroleum oil, household insecticide, 35
Phenothrin, household insecticide, 33
Photodegradation, metalaxyl, 5, 7
Photoproducts, metalaxyl, 7
Pindone, household rodenticide, 36
Pine oil, household insecticide, 35
Piperonyl butoxide, household insecticide, 32
Propoxur, household insecticide, 32
Proton NMR spectroscopy, toxicological biomarker, 118
Pyrethrin, household insecticide, 32
Pyrethroid household insecticides, 32
Pyriproxyfen, household insecticide, 34

Q^*_1 (oral cancer slope factor) values, household pesticides, 47, 50

RCRA (Resource Conservation and Recovery Act), 27
Repellents, household pesticides, 35
Resmethrin, household insecticide, 33
Resource Conservation and Recovery Act (RCRA), 27
RfD values, household pesticides, 47, 50
Risk, household pesticide disposal landfills, 38
Rodenticides, household pesticides, 36

Soil half-lives, household pesticides, 31
Soil health measurements, recommendations, 127
Soil invertebrate biomarkers, limitations, 130
Soil invertebrate biomarkers, overview (table), 132

Soil invertebrate biomarkers, potential, 129
Soil invertebrate sampling methods, 128
Soil pollution, terrestrial invertebrate biomarkers, 93 ff.
Solid waste landfills, improved standards, 57
Stress protein families, characterized, 97
Stress proteins, invertebrate biomarkers, 97
Stress proteins, low molecular weight, 101
Stress-60 (Chaperonin, hsp60) proteins, biomarkers, 100
Stress-70 (hsp70) proteins, biomarkers, 99
Stress-90 (hsp90) proteins, biomarkers, 99
Sulfluramid, household insecticide, 34
Sumithrin, household insecticide, 33

Terrestrial invertebrate biomarkers, 93 ff.
Terrestrial invertebrates, soil risk assessment, 93 ff.

Tetrachlorvinphos, household insecticide, 31
Tetramethrin, household insecticide, 33
Thin layer chromatography, metalaxyl, 11
Toxicity category definitions, household pesticides, 53
Toxicity, household pesticides technical materials, 50
Tralomethrin, household insecticide, 34
Trichlorfon, household insecticide, 31

Ubiquitin, stress protein, 98
Ultrastructure, invertebrate, as biomarker, 109

Warfarin, household rodenticide, 36

Z-9 tricosene, household insecticide, 35
Zinc phosphide, household rodenticide, 36